T0093597

MARS

Kosmos

A series exploring our expanding knowledge of the cosmos through science and technology and investigating historical, contemporary and future developments as well as providing guidance for all those interested in astronomy.

Series Editor: Peter Morris

Already published:

Jupiter William Sheehan and Thomas Hockey
Mars Stephen James O'Meara
Mercury William Sheehan
The Moon Bill Leatherbarrow
Saturn William Sheehan
The Sun Leon Golub and Jay M. Pasachoff

Mars

Stephen James O'Meara

REAKTION BOOKS

To Deborah, my Princess of Mars

Published by Reaktion Books Ltd
Unit 32, Waterside
44–48 Wharf Road
London N1 7UX, UK
www.reaktionbooks.co.uk

First published 2020

Printer and bound in India by Replika Press Pvt. Ltd

A catalogue record for this book is available from the British Library

ISBN 978 1 78914 220 4

CONTENTS

ONE

Communion with Mars

Mars has burned its imprint on the human imagination ever since stargazers first pondered its appearance in the night sky. This blood red 'star', wandering among the fixed stars like a wounded animal, probably first invoked fear, then wonder, in our ancestors, especially with the emergence of anthropocentric thinking. Eventually they conferred upon Mars some larger purpose related to human destiny. These thoughts evolved into the practice of astrology; the appearance of Mars in different constellations, the length of its stay and its brightness variations – combined with the planet's erratic motions, unions with the Moon and planets and the angles between them, among other things – offered astrologers endless possibilities for employing the planet in their personal deductions, as exemplified in this ancient Assyrian text: 'When Mars is dim, it is lucky; when bright, unlucky. When Mars follows Jupiter that year will be lucky.'[1]

As the ancients sought to position themselves in their perceived cosmos, the rise and fall of Mars's light – mimicking, over a much longer timescale, a pulsing heart – may have also taken on a spiritual significance, the first possible traces of which we find haunting their caves as paintings. Africa's Basarwa people gave birth to the earliest-known art rock drawings some 70,000 years ago in Botswana's Tsodilo Hills. (This UNESCO

True colour image of Mars taken by the OSIRIS instrument on the European Space Agency (ESA) *Rosetta* spacecraft during its February 2007 flyby of the planet.

World Heritage Site has more than 3,500 paintings, the largest concentration in the world, with some more than 20,000 years old.)

The Basarwa were also quite possibly the first known to think abstractly and behave like modern people.[2] According to the Basarwa creation myth, humankind descended from a python in the sky. A symbolic representation of the sacred reptile, one of their most revered animals, appears as a series of ritual indentations in a megalithic rock projecting out of a cave in the Tsodilo Hills that trace the serpent's form. As the python was from the sky, the indentations representing the snake may be one of the earliest records of the Milky Way; the serpent has represented the band of the Milky Way in later cultures all across the globe, and is arguably the most widespread mythology known to humanity. The passage of Mars through the python (as it does during its most favourable apparitions, when the planet is at its brightest and directly overhead in the Southern Hemisphere) would not have gone unnoticed, as the crimson planet (the colour of kill) coiled through its stars like an animal in the python's grasp.

A series of ritual indentations in a megalithic rock projecting out of a cave in the Tsodilo Hills, Botswana, may represent the mythical python from which humankind descended, according to Basarwa belief.

Mars overhead in the Botswana night sky in August 2016. The planet is on its way towards being captured by the celestial python (the Milky Way).

While we can only guess how the Basarwa interpreted Mars, the above theory makes sense as the early people had similar mythologies related to the Sun and Moon: the Sun is an elephant that is hunted and killed every day at sunset (giving the Sun its blood-red colour when near the horizon); likewise the Moon is an eland (a type of antelope) that suffers the same fate. And whenever the Moon appeared extremely red, such as during a lunar eclipse, it was a sign that someone had died or that some great animal had succumbed to the hunt.[3] A union of Mars and a totally eclipsed Moon over Botswana in 2018 carried more religious overtones to one ill-informed side-of-the-road preacher, who presented this natural event as a supernatural display of Satan's power to transform good into evil – only to end when the higher power answered determined prayers to stop the event.[4]

A blood moon together with blood-red Mars during the 27–28 July 2018 total lunar eclipse over a Botswana landscape with elephants.

Shaman during a trance dance in Botswana.

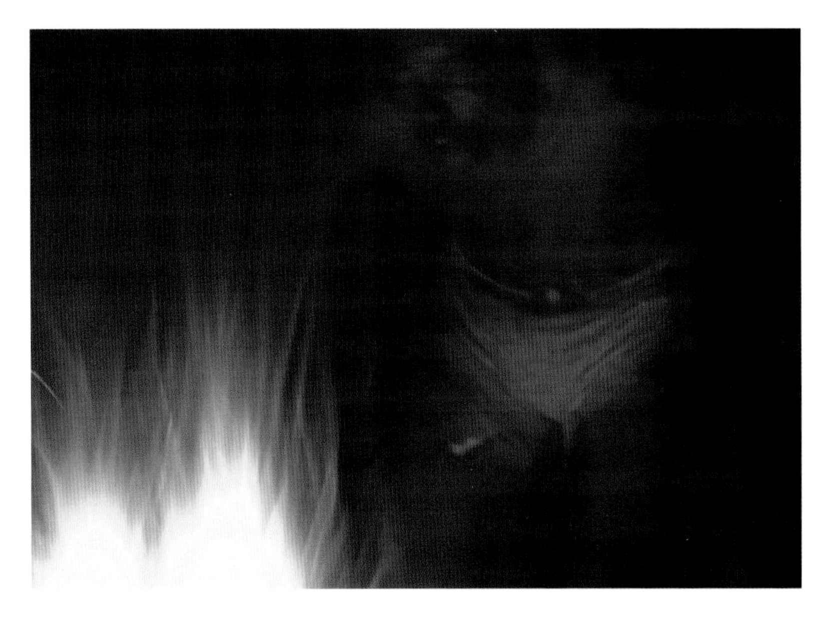

The transformation of a celestial body (like Mars) into an animal was important to the early Basarwa's spiritual beliefs, some of which we still see practised today in a certain trance dance performed around the time of full Moon. As the shaman dances, stamping rhythmically to rattle silk-moth-cocoon bracelets tied around his ankles, in a circle around a fire, women sitting in an outer circle around the fire clap and chant to help activate the spiritual energy surging through him. The dance climaxes when the shaman feels as if he has transformed into an animal and can communicate with the spirit world.[5]

Early Basarwa believed that their ancestors (or important animals) lie among the stars. The brighter the star (or planet) the more prominent the person or animal (and the more recent the death). To some of today's Basarwa, Mars serves a more utilitarian significance, sharing duties with other bright stars and planets as a fire-starting star: when the red orb rises in the east after sunset, it is time to make a fire; when the planet sets in the west before sunrise, it is time to extinguish the fire.[6]

Mars was also part of the Australian Aborigines' Dreamtime – a vision from a time beyond memory, a mystical part of the culture that, as with the early people of Africa, has been handed down by legends, songs and dance for more than 40,000 years. To the Kamilaroi and Euahlayi people of New South Wales, the ruddy colour of Mars held special significance, being a key player in their Morning Star Ceremony to ensure that the deceased travel safely to the Land of the Dead. The rising of Mars in the east after sunset signalled the time to light the ceremony's Sacred Fire, which remained burning until Mars set in the west around the time of sunrise,[7] a ceremony strikingly similar to the Basarwa's more functional practice described above.

Fall of an Aboriginal Mars Legend?

For more than 150 years a tale was shared that the Coorong of South Australia knew Mars as *Waiyungari* (also *Waijungari* and *Wyungari*), a newly initiated man who was ceremonially covered with red ochre. According to legend, Nepeli, a peer of the Supreme Being, caught Waiyungari engaging in illicit behaviour with his two wives. To escape punishment, Waiyungari hurled spear after spear (one into the other) into the night sky, creating a shaft long enough for him and the two women to use to climb into the Milky Way, where they would be safe from the vengeful husband.[8]

In 2017, however, Duane W. Hamacher re-analysed the traditions and says that anthropologists appear to have misidentified Waiyungari as Mars, and the two women as other planets. Hamacher counters that Waiyungari is the red giant star Alpha Scorpii (Antares; the rival of Mars in Roman tradition), and Tau and Sigma Scorpii, which flank Antares, are the two women. They are in the aboriginal constellation known as the 'celestial canoe' (the stars of Scorpius), which lies in the river Milky Way.[9]

'He Travels Backwards'

As early as the second millennium BC the Babylonians bestowed several human qualities on Mars that were either adopted or independently recognized by later cultures. They primarily recognized Mars as their great hero Nergal, god of the Underworld and bringer of plague, war and famine. A key player in astrological predictions, Nergal was believed to have the power of prophecy.[10] Omen texts began to relate the phenomena of Mars to predictions about military success and failure. In Babylonian astrology the planets possessed an intimate relationship with the moment of birth, which helped define the person's character and quality of life, such as 'a child born under Mars may have a hot temper'.[11]

The Babylonians may have been the first to link Mars the wanderer to agricultural concerns, calling it, at times, *Apin* (the

A new interpretation of an old aboriginal legend claims that Antares is the newly initiated man Waiyungari with two women, represented by Tau and Sigma Scorpii.

The apparent retrograde (backward) motion of Mars against the stars.

plough), suggesting they recognized the planet's golden colour when closest to Earth. Mars then had two faces: one as a god of war; the other as a god of agriculture. They also recorded Mars's colour (*Makrû*, meaning Red Star), which they associated with fire, lending it a secondary function as the Babylonian patron of magic and forging.

Mesopotamian skywatchers were well aware that something special happens to Mars when it appears opposite the Sun in the sky, a position known as *opposition*. For most of Mars's passage it appears as a moderately bright star, about as bright as any in the Big Dipper or Plough, moving west to east against the fixed stars. Around the time of opposition, however, just about the time when Mars reaches its peak brightness, rivalling brilliant Jupiter or Venus, the planet performs an apparent loop: it stops moving forward, reverses direction for a short time, stops again, resumes its forward motion, then gradually fades to invisibility as it approaches the Sun.

The Babylonians may have recognized this motion, as they also knew Mars as *kakkab la minâti* (unpredictable one), suggesting perhaps that Mars's backward motion caused confusion in their predictions of where the planet should appear.[12] The Egyptians, however, were the first to record it, saying in an epithet about Mars, 'he travels backwards'.[13] It may also be recorded in the earliest-known Egyptian star map, on the ceiling of the tomb of Senenmut – the chief architect and astronomer during the reign of Queen Hatshepsut (r. c. 1473–1458 BC) – at Deir el-Bahari on the west bank of the Nile opposite Luxor. There we see Mars depicted as an empty boat in the west. The map shows the other naked-eye planets as boats as well, but with figures standing in them. This special treatment of Mars has been interpreted as a record of its odd behaviour in the sky – namely the planet's profound ability to move forward, reverse and stand still, before it moves forward again along the zodiac.[14]

The Egyptians originally knew Mars as 'Horus of the Horizon' or 'Horus the Red', suggesting it represented an aspect of the Sun

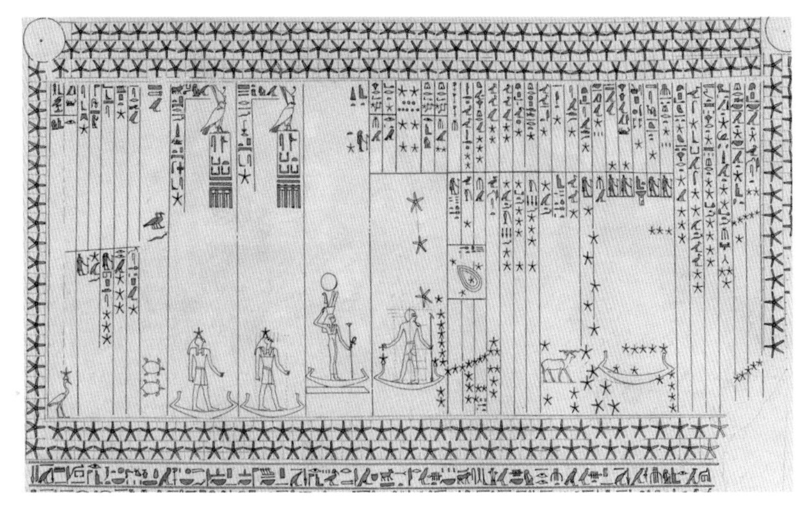

The southern part of the astronomical ceiling in Senenmut's tomb, showing the unoccupied boat of Mars at lower right, while the other naked-eye planets appear as boats with figures standing in them.

god when it turns red near the horizon. This further promotes the Babylonian tradition of bestowing deity status on the planet; the Egyptians later singled out Mars for its colour, simply referring to it as *Har Decher* (the Red One).[15]

As with the Babylonians, early Roman stargazers noticed the dual nature of Mars. While the Red Planet is popularly known today as the Roman god of war, the Italian Mars originally had the surname Silvanus; as a divinity of the forests and protector of herds and cattle, as well as a god who watched over the fields and protected their boundaries, citizens offered sacrifices to him for the prosperity of their fields and flocks.[16]

The warlike Mars had the surname Gradivus, probably derived from *gradior*, meaning 'to march'. A temple dedicated to Mars Gradivus was located beside the Appian Way, Rome, outside the Porta Capena, which was a gate in the Servian Wall near the Caelian Hill; soldiers sometimes halted near his temple when they marched out to battle. Mars then morphed into different forms corresponding to the successive conditions of the Roman citizen: first a farmer (when golden Mars stood close to Earth to guard the crops), then a bloody warrior (when the planet marched away from Earth, turning

Left: Red Mars in January 2018 when the planet was about 280 million km from Earth, as viewed from the southern hemisphere.

Right: Golden Mars in June 2018 when the planet was about 75 million km from Earth (with ample dust brewing in its atmosphere). Mars was much brighter and larger in June than in January, but the images here have been scaled for colour comparison.

blood red as it went off to protect the nation in battle). 'The transition from the idea of Mars as an agricultural god to that of a warlike being, was not difficult with the early Latins,' according to a contributor to a *Dictionary of Greek and Roman Biography and Mythology*, 'as the occupations were intimately tied.'[17]

The Era of Explanation

By the sixth century BC the Greeks began to look at the heavens with a yearning to understand it through logical enquiry, including Mars's still mysterious retrograde motion. Eudoxus of Cnidus (390–337 BC), a Greek mathematician and astronomer and pupil of Plato (428–348 BC), was the first to arrive at a solution. He designed a theoretical model of the solar system using multiple concentric spheres and supposed every planet to be affixed to a sphere revolving around the Earth as centre.

In this model, which he called the 'principle of uniform planetary motion', Mars occupied the fifth sphere. The fixed stars occupied the outermost sphere. He then tilted the axis of the sphere containing Mars and made it rotate in the opposite direction of the outer sphere of stars. He kept the period of rotation the same for both. The combined uniform motion of these two spheres resembles what is known in mathematics as a *hippopede*, namely, a figure of eight. Finally, to make Mars circle the Earth, he had the two spheres rotate about a third sphere whose axis is roughly perpendicular to that of the outer sphere. The result is that the figure of eight is smeared across the sky – predominantly west to east, but occasionally east to west, as we see in the retrograde motion of Mars.

Eudoxus envisioned the spheres only as mathematical devices to explain the observed motions. Brilliant as the model was, explaining to some degree the retrograde motion of Mars, it did not explain

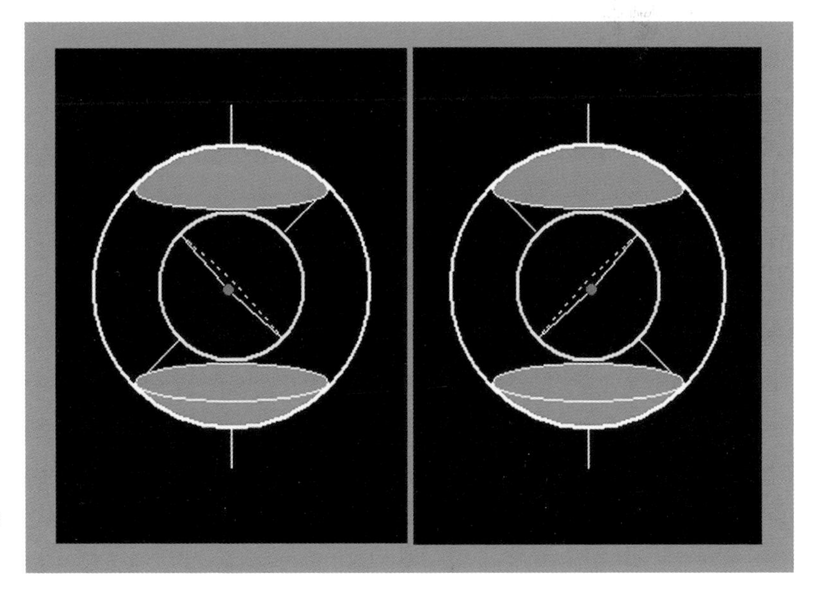

Eudoxus used two nested spheres to model the 'principle of uniform planetary motion', like a compass on a gimbal. Mars rotates in a tilted inner sphere at the same speed but opposite to that of a vertical outer sphere, causing us to see Mars (fixed to the equator of the inner sphere) move in a figure-of-eight pattern.

why the planet varied in brightness; if Mars moved around the Earth in a circle, its brightness should not vary at all.

It took the imaginings of Claudius Ptolemy (*c.* AD 100– *c.* 170) – the 'pro-astrological authority of the highest magnitude' – of Alexandria, Egypt, to bring Mars's motions to a near-perfect state.[18] In Ptolemy's model, Mars travelled on a small circle (*epicycle*) that moved around the Earth on a larger circle (the *deferent*). As the epicycle turned, a point on its rim followed a loop path, swinging in towards the centre of the deferent, before moving outward again in reverse, just as called for to explain the retrograde motion of the planet.

But Mars does not display a uniform speed in its orbit; when the planet comes to opposition, it moves twice as fast as it does when on the other side of its orbit. To solve this problem, Ptolemy made his deferent circle off-centre from the Earth. This solution took care of not only the planet's variations in speed but its differences in brightness (being brightest when closest to the Earth and dimmest when farthest away).

Graeco-Roman astronomer-astrologer Claudius Ptolemy (*c.* AD 100–*c.* 170).

The only mystery left: how to explain the different sizes of the retrograde loops. For this, Ptolemy made Mars move uniformly – not around the centre of the deferent but around another point called the *equant*, which is as far from the centre of the deferent as the Earth. One beautiful outcome of Ptolemy's theory was ensuring that the retrograde loop of Mars would always occur when Mars is opposite the Sun, which it does.

Ptolemy's model reigned until the early 1500s. Sometime between the latter half of 1508 and early 1514, Polish astronomer Nicolaus Copernicus (1473–1543) began

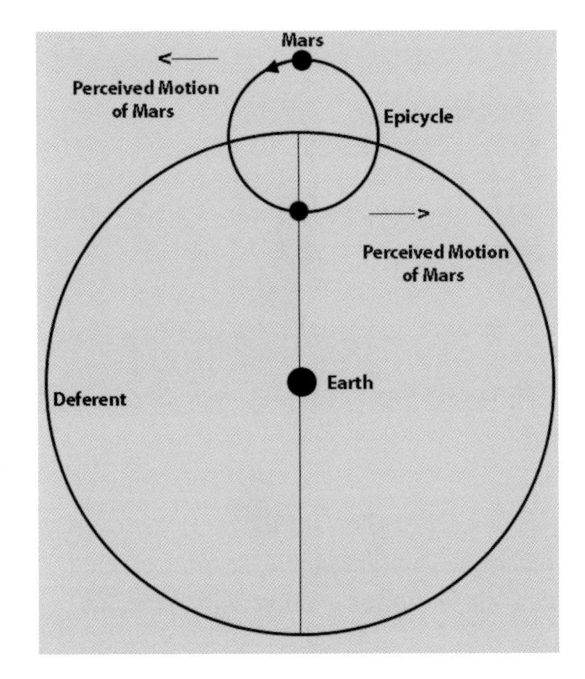

Ptolemy's model (in simplest form) explaining retrograde motion, incorporating Mars travelling on a small circle (the epicycle), around a larger circle (the deferent).

Below: As Mars moved in its orbit it traced a retrograde loop among the stars with two stationary points.

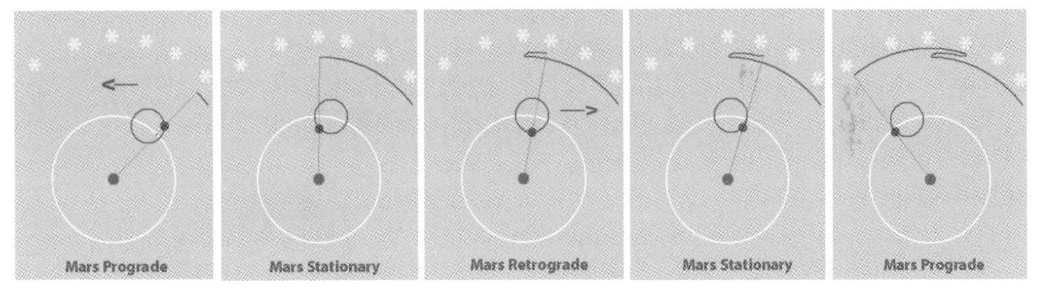

circulating among a few trusted professional friends a short hand-written sketch, the *Commentariolus* (Little Commentary), suggesting a new planetary theory, the roots of which stretched back seventeen centuries to a Sun-centred system first proposed by Aristarchus of Samos (c. 310–c. 230 BC).[19]

Copernicus argued that the Earth is an ordinary planet in orbit around the Sun. He suggested that those who believed in the Earth-centred theory before him must have suffered the same illusion that people on board ship have while moving through quiet

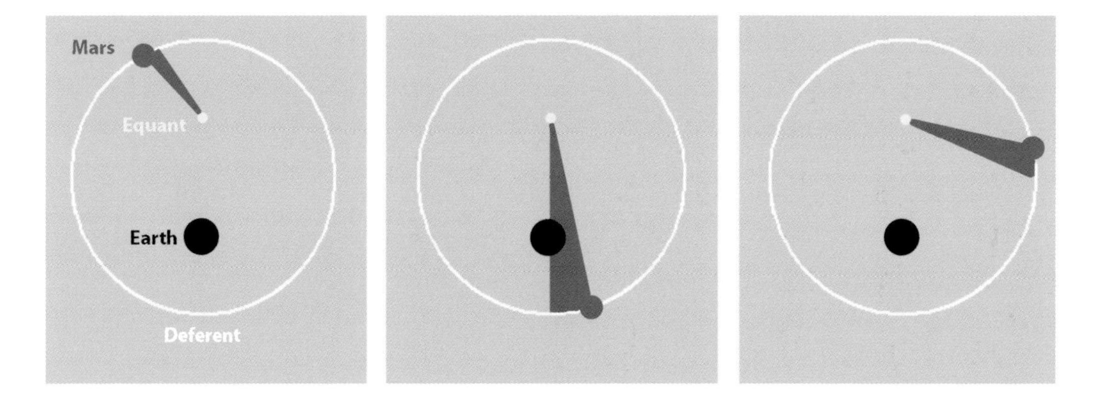

Ptolemy's solution to not only the varying speed and brightness of Mars (by removing the Earth from the centre of the deferent), but the different sizes of the retrograde loops (by having Mars move uniformly around another point called the equant).

water; they would think the vessel at rest while houses and trees on shore seemed to be in motion.[20] He also believed that Ptolemy's introduction of the equant, a mere mathematical device, could not have a real physical significance. How simple and natural it would be, he said, for the Earth to circle the Sun – along with Mars and the other planets.

So Copernicus began work on a Sun-centred model of the universe, culminating with the publication in 1543 of his book *De Revolutionibus orbium coelestium* (On the Revolutions of the Celestial Spheres). In this he correctly proposed that the retrograde motion of Mars is an optical illusion, explaining that as the Earth rounds the Sun (twice as fast as Mars), we see Mars move backwards due to a shift in perspective. The idea was brilliant.

Nevertheless, there arose a wave of opposition from Protestant theologians, who regarded *De Revolutionibus* as heresy, but several decades would pass before the Catholic Church turned against the theories of this renowned astronomer and respected administrator in minor orders – Copernicus remained a Catholic – and persuaded him to ban his Sun-centred views in the early seventeenth

Nicolaus Copernicus (AD 1473–1543); the man who shattered Ptolemy's Universe.

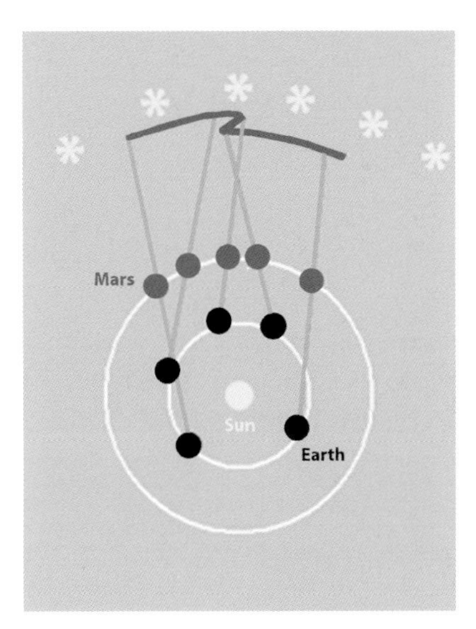

As the Earth's orbital motion carries it around the Sun counter-clockwise nearly twice as fast as Mars, our perspective of the planet changes, causing the illusion of retrograde motion.

century.[21] While it has long been alleged that Martin Luther was 'one of the prominent voices of the Protestant opposition to Copernicus's ideas', historian of science Andreas Kleinert, from Martin Luther University in Halle, Germany, has shown that 'the famous citation from Luther's table talks [claiming that Copernicus is a "fool who would overturn the science of astronomy"] is next to worthless as an historical source, and that there is no indication that Luther ever suppressed the Copernican viewpoint.'[22] Like the Greeks, however, Copernicus was unable to conceive of any motion other than circular for a planet, which failed to explain why retrograde loops of Mars were not always the same. He died unaware of the solution.

The resistance to the Copernican theory was not limited to theological considerations. Tycho Brahe (1546–1601), a Danish nobleman and leading observer in the generation after Copernicus, was frankly sceptical of it. The cause may arguably be either due to his astrological belief that heavenly bodies influenced terrestrial events, or, as a deeply religious man, his literal interpretation of the Bible – namely, that the Sun must be the moving body, for Joshua ordered the Sun to stand still and not the Earth. Although his

Tycho Brahe (1546–1601). From the 1655 edition of *Tychonis Brahei Equitis Dani* by Pedro Gassendo (Pierre Gassendi).

observational skills were indisputable, and his dedication to mathematical precision acute, Tycho still clung to some mystical traditions, such as accepting the powers of both astrology and alchemy. Superstition steered his life: he would discontinue a journey, for instance, if he saw a hare or met an old woman.[23]

Tycho Brahe's 'astrologically directed mind' chiefly focused on the planets – and Mars in particular. The Red Planet fuelled an obsession that led him to observe every opposition from 1580 until his death in Prague in 1601 (the last of these being in 1595). In 1583 Tycho noted that near opposition Mars was moving retrograde at a rate of nearly half a degree every day. This proved that the planet could approach much nearer to the Earth than the Sun, which was possible in the Copernican system, but impossible in the Ptolemaic. Nevertheless, Tycho could still not bring himself to accept the ideas of Copernicus, and instead adopted a compromise position – the Tychonian system – in which the Earth remained at the centre of the universe, the planets circled the Sun, while the Sun in turn moved around the Earth.

A Gateway to a 'New Astronomy'

The year before his death, Tycho was joined by Johannes Kepler (1571–1630), the German-born mystic, mathematician and confirmed Copernican. Together they worked on the theory of the motions of Mars. In September 1601, when Tycho had only a month to live, Kepler began what he called his 'warfare on Mars'.[24] Using Tycho's precise data on the positions of Mars, the stars and the other planets, Kepler deduced, through punishing calculations, three laws of planetary motions that solved the mystery of Mars. First, Mars does not move around the Sun in a circular orbit, but in an ellipse (a large oval that carries the planet closer to the Sun at certain times than at others). Second, Mars travels fastest when near the Sun, and slowest when farthest away. And finally, he linked the time it took Mars to orbit the Sun with its distances from the Sun, which held the key to working out the scale of the solar system, as these three laws (now known as Kepler's Laws of Planetary Motion) hold true for all the other planets.

With these laws, Kepler succeeded in sweeping away the vestiges of Ptolemaic astronomy, ending what Harvard astronomical historian Owen Gingerich called 'the two-millennia spell of perfect circles and uniform angular motion'.[25] His work opened a gateway to the 'New Astronomy' looming on the near horizon. In 1609, the same year that Kepler published his results on the elliptical orbit of Mars, news broke that a new instrument for drawing faraway objects closer had been invented in Holland. The telescope, as it was called, was soon at the disposal of astronomers and Kepler was the first, in 1611, to present a theory of the course of light rays through the lenses and of the formation of an image.

TWO

The 'Miniature of Our Earth'

Tycho Brahe's precise measurements of planetary positions, Copernicus's new theory of how the planets orbit the Sun and Kepler's three laws of planetary motion introduced a new era of Mars observation. In 1609 Galileo Galilei (1564–1642), a professor of mathematics at the University of Padua, turned a 20× telescope of his own making towards the planet but saw only a tiny orb of light. While the view must have come as a disappointment to him – especially after seeing craters on the Moon, the phases of Venus and four moons orbiting Jupiter through his new telescope – this result is not surprising. Mars's diameter is only half that of the Earth's and never appears any larger than one-hundredth of the Moon's apparent diameter (or about the size of an average crater on the Moon). Seeing decent surface details on the planet's surface requires telescopes of larger aperture and greater magnification than he was using.

Galileo, however, ultimately found ways to master his views of Mars. He noticed, for instance, that when Mars was viewed with the unaided eyes at perigee (when the planet is closest to Earth) its light 'greatly surpasses Jupiter in brightness'; but turn a telescope to Mars and its disc will appear greatly inferior to that of Jupiter, revealing that Mars, 'in actual size [is] far inferior to Jupiter'.[1] He also noticed that Mars was not completely circular when compared to Jupiter; as he wrote in a letter dated 30 December 1610 to a former pupil,

Benedetto Castelli: 'I ought not to claim that I can see the phases of Mars; however, unless I am deceiving myself, I believe that I have already seen that it is not perfectly round.'[2]

Nearly three decades would pass before telescopes advanced enough to show detail on the planet. As early as 1638 the Neapolitan scholar Francesco Fontana (1580–1656) noticed clearly the gibbous phase of Mars, beginning what the French astronomy popularizer Camille Flammarion (1842–1925) described as the first period of the history of Mars. Fontana also drew Mars with a 'black cone' like a hollow in the middle of the planet. In a 2017 article Paolo Molaro provides evidence that it was probably the dark Syrtis Major marking on Mars, the most prominent feature visible on the disc to this day. The dark spot changed quickly and, from its quick motion, Fontana deduced that Mars revolves around its own axis.[3]

Fontana also recorded a dark circular band surrounded by a brighter one at the limb. Although this feature has largely been discounted as either an illusion or telescopic defect, I believe the bright circle may have been his way of recording limb brightening, while the dark band was a simultaneous contrast illusion.

Further observations of surface markings on Mars were slow coming: on 24 December 1644 the Jesuit Daniello Bartoli (1608–1685) at Naples saw Mars almost round with two patches above

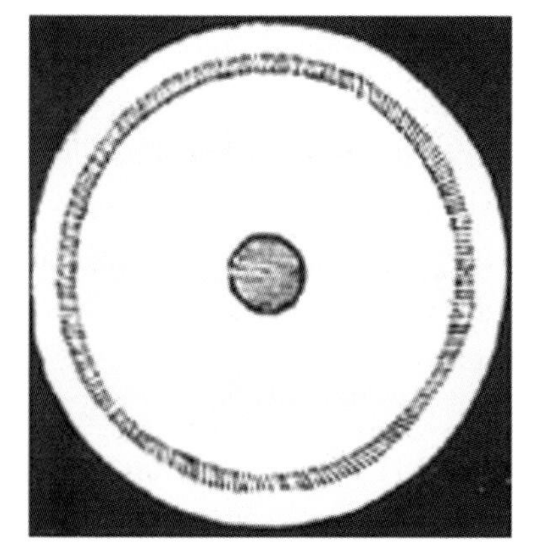

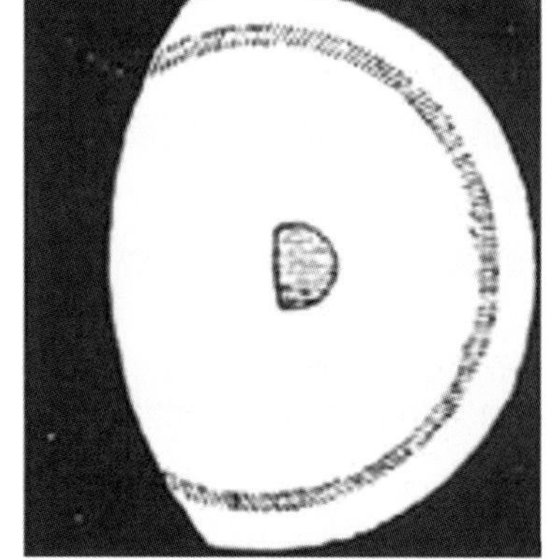

Francesco Fontana's drawings of Mars: full disc, 1636 (left); and in gibbous phase, August 1638 (right).

Drawing of Mars by Christiaan Huygens made on 28 November 1659, showing the distinct V-shaped marking known as Syrtis Major.

the centre of the disc; and in 1651 Giovanni Battista Riccioli (1598–1671) and Francesco Grimaldi (1618–1663), two Jesuits of the Collegio Romano in Rome, recorded still more ill-defined dark patches of varying reflectivity. But it was the Dutch astronomer Christiaan Huygens (1629–1695) who, on the evening of 28 November 1659, included in his drawing of Mars an indisputable recording of the V-shaped Syrtis Major dark surface marking, giving birth to the study of Martian geography.

Part of Huygens's success was due to his more powerful telescope (7 cm aperture with a focal length of 3.2 that produced a magnification of 87×). The enlarged disc helped him to notice a slight shift in the marking against the planet, over the course of two and a half hours, indicating its rotation. Three nights later, the marking was nearly in the same location on the disc. Seeing that the patch was, once again, shifting position over the course of his observation (much too quickly for it to be a three-day period), he wrote in his journal on 1 December 1659: 'The rotation of Mars, like that of the Earth, seems to have a period of 24 hours.'[4]

In 1666 Huygen's rival, the Italian astronomer Giovanni Domenico Cassini (1625–1712), independently studied Mars at the Panzano Observatory in Bologna, where he made about twenty crude drawings of the planet that showed various markings on the disc. The features, he noticed, returned to the same positions about forty minutes later than on the previous day. Moreover, after a period of 36 or 37 days, they returned to precisely the same positions at the same hour of night. This led Cassini to refine Huygens's rotation period of Mars to 24 hours 40 minutes (a figure less than three minutes off from the planet's true rotation period of 24 hours 37 minutes 22.66 seconds).

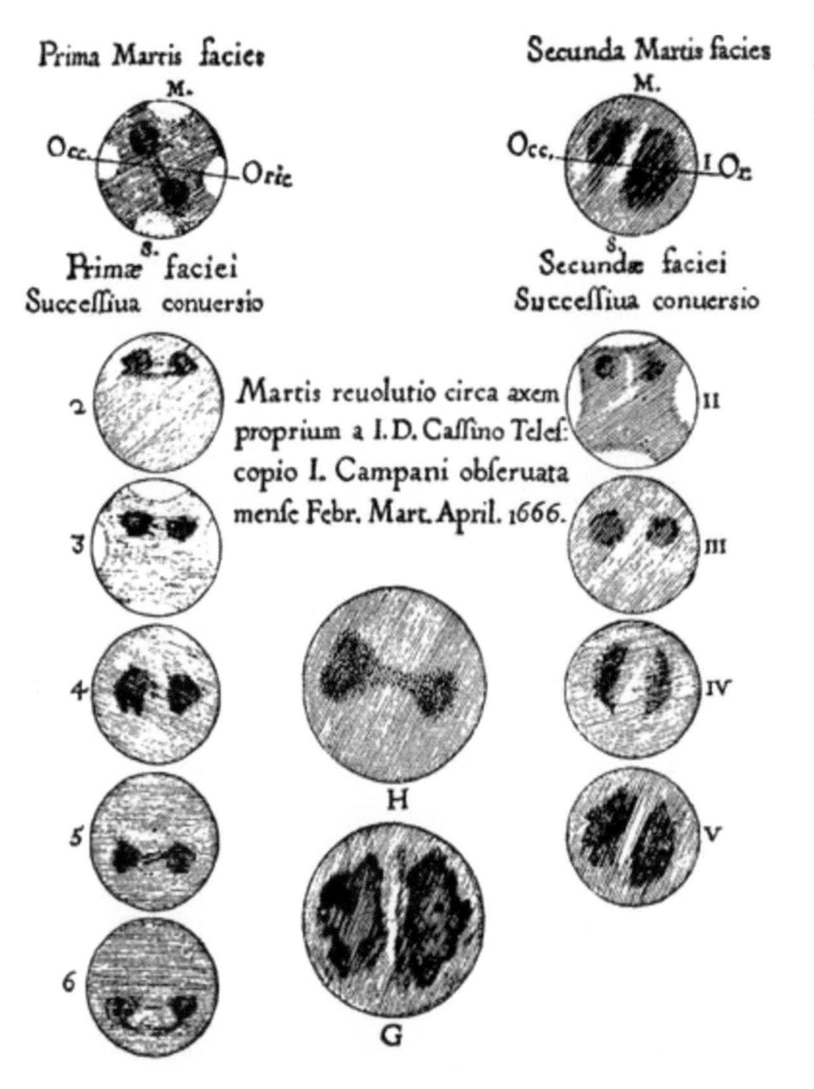

Drawings by Giovanni Cassini, made in spring 1666, showing surface features on Mars.

While Huygens initially doubted Cassini's claim, he later came to terms with it, establishing it as fact in his *Cosmotheoros*, published posthumously in 1698; in this work Huygens speculates about intelligent extraterrestrials, making it the first published exposition of extraterrestrial life.[5] Another of Cassini's discoveries in 1666 had been one of the polar caps of Mars, which Huygens independently observed in 1672. The planet's geography was expanding, making Mars appear ever more a mirror of the Earth.

In his *De Planetarum maculis* Cassini twisted the scenario around, in essence reasoning that if we were to view the Earth from a great distance across space, our world would resemble Mars: Earth's seas would appear dark because they absorb sunlight, while the continents would appear bright. To this day we still refer to the dark Martian surface features as 'mare' (meaning sea) and its ochre sands by Latin continental names, even though they are misnomers.

During the favourable apparition of Mars in 1672, Cassini set out to determine just how great is the gulf of space that separates Earth from Mars. Using his own observations from Paris (one point of a triangle) and those by Jean Richer (1630–1696) in Cayenne, French Guiana (the second point of a triangle), Cassini triangulated the distance of Mars (the third point in a triangle); in other words, by sighting Mars from two different locations on Earth, and knowing the distance between those locations (7,080 kilometres (4,400 mi.)), Cassini could use trigonometry to determine the distance to Mars. His value of the astronomical unit (the average distance from the

The Paris Observatory and its 10.3 m aerial telescope; depiction probably by Charles-Joseph-Étienne Wolf (1827–1918).

Earth to the Sun) was only 7 per cent short of that accepted today and gave the first accurate indication of the scale of the solar system.

Studying Martian Geography

Cassini's nephew, Giacomo Filippo Maraldi (1665–1729), carried out the best work on Mars of the next generation. During the favourable 1704 opposition, he used the 10.3-metre (33.8 ft) focal length aerial telescope at Paris Observatory to observe a dark band on Mars with a small protrusion, which made it easy to identify. Over time, he noticed the features on Mars changing shape from one month to the next. 'Notwithstanding these changes,' he said, 'the patches last long enough for us to follow them for a time sufficient for a determination of the rotation period.'[6] Richard McKim, Mars Section Director of the British Astronomical Association, suspects that the reason for the changes may have been the presence of a dust storm in the area. If true, Maraldi's observation is the first telescopic record of such an event on Mars.[7]

Maraldi made several more remarkable observations of Mars during its very close 1719 opposition, when the planet appeared so brilliant in the sky that the public mistook it for a new star or comet stained with blood. Most noteworthy was his studies of the white spots at the Martian poles, which displayed apparent fluctuations: in the Martian winter, the southern spot (which was then in view) suffered slight diminution as Martian summer approached – before receding rapidly until it finally vanished, only to regrow once again. Unable to explain the polar phenomena, which he linked to the rotation of the planet, Maraldi likened them to sunspots or the cloud bands on Jupiter.[8]

It would take the genius of the German-born, English astronomer William Herschel (1738–1822) to correctly deduce the answer. In 1777 he began a systematic observation of Mars, using a 23-centimetre home-made reflecting telescope (based on the design

Edward Scriven, after John Russell, *Sir William Herschel*, engraving, early 19th century.

introduced by Sir Isaac Newton in 1672) in the south-facing garden behind his house in Bath. Using a magnification of 211×, Herschel paid particular attention to Maraldi's white polar spots and the planet's rotation. After studying four successive oppositions of Mars, Herschel explained the spots as most probably caps made up of water ice: 'the bright polar spots are owing to the vivid reflection of light from frozen regions; and that the reduction of those spots is to be ascribed to their being exposed to the Sun.'[9]

Herschel also noticed that the vanishing spot of summer was slightly offset from the rotation axis of Mars, concluding that the point of most intense cold must lie some distance from the Martian geographical pole, as in the case with Earth. (We now know that his suspicion is correct, as measurements by the *Mars Global Surveyor* spacecraft have placed the cap's greatest elevation some 250 kilometres (400 mi.) from the pole.) In addition, he determined that the planet's axis of rotation was only slightly more tilted than the Earth's (24° axial tilt for Mars; 23½° for Earth), which meant seasons on Mars would be much like ours, although twice as long. Herschel concluded that the inhabitants of Mars 'probably enjoy a situation in many respects similar to ours'. This statement was not only the first unequivocal one of Mars's likeness to the Earth, but one that ignited the imagination with possibilities of life beyond the Earth.

Mars studies continued with Johann Hieronymus Schröter (1745–1816), an enthusiastic amateur astronomer who stepped down from an important post with the Elector of Hanover (the British King George III) to serve as chief magistrate at the village of Lilienthal, near Bremen. There, in 1781, he constructed the largest observatory in Europe, equipped with the largest telescope on the Continent. His observations there, until the observatory was wrecked by French troops in 1813, convinced him that the various light and dark areas on the planet were the changing forms of a mere 'shell of cloud', a view apparently shared by the French astronomer Honoré Flaugergues, who observed Mars at his private observatory at Viviers between 1796 and 1813. While some of these changes followed the seasons, it has also been proposed that, like the earlier observations of Maraldi, these changing forms may have also been due to dust activity.[10]

A new dimension to Mars studies, Martian cartography, began in the 1830s with Berlin astronomers Wilhelm Beer (1797–1850) and Johann Mädler (1794–1874). They combined results from ten years of observations to make the first map of Mars in 1840, which showed its more consistent features. The observations were made using a 9.5-centimetre (3.7 in.) refracting telescope, which achieved a magnification rate of 185×, resulting in beautifully sharp and well-defined images far superior to those used by previous observers.

Perhaps influenced by Cassini's nomenclature, Beer and Mädler found resemblances between the Martian forms and the Earth's

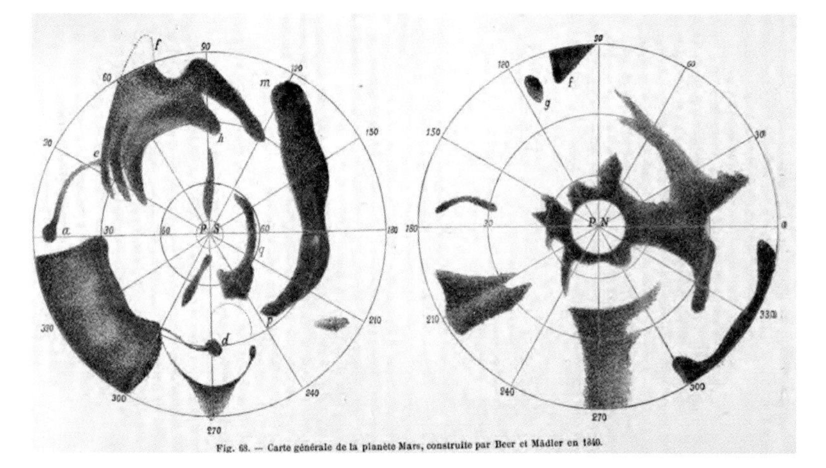
Fig. 68. — Carte générale de la planète Mars, construite par Beer et Mädler en 1840.

Maps of Mars created in 1840 by Wilhelm Beer and J. H. Mädler – the first geographical charts drawn of the Martian world, they became the classical reference for all later observers.

oceans and lands. The feature they found most characteristic of the planet was a small round patch, which appeared to be dangling from an undulating ribbon. Although Herschel had detected this feature as early as 1783, Beer and Mädler were the first to represent it clearly. The spot was 'so near the assumed equator' that they chose it as the 'reference point for the determination of the planet's rotation'. Beer and Mädler's work, the first methodical attempt at studying Martian geography, revealed a rotation period of 24 hours 37 minutes 23.7 seconds – a figure within one second of its true value.

The cartographers also used this dark patch, known today as Sinus Sabaeus–Meridiani, to mark the planet's prime meridian (zero-degree line of longitude, the Martian equivalent of Earth's Greenwich) before they plotted other fixed features onto a geographic grid they fashioned for the planet, which would be adopted by all later astronomers. Many of the features they rendered are still identifiable today, including the dark 'V'-shaped Syrtis Major, the oval Solis Lacus (Eye of Mars) and the Trapazoidal Mare Acidalium. According to Flammarion, Beer and Mädler deserve to be remembered as the 'true pioneers in this new conquest of Mars'.[11]

The notion of Mars as a mirror of Earth continued with Jesuit astronomer Father Angelo Secchi (1818–1878), who noted the resemblances found between the Martian forms and Earth's oceans and lands, commenting that 'Everything is variegated like a map of the Earth.' Flammarion further promoted this view of Martian geology in 1873, painting the Red Planet as a remarkable facsimile of Earth, though with dissimilarities as well:

> On our planet the seas have greater extent than the continents . . . It is different with the surface of Mars, where there is more land than sea, and where the continents, instead of being islands emerging from the liquid element, seem rather to make the oceans mere inland seas – genuine mediterraneans . . . The

> seas and the straits which connect them constitute a very distinctive character of Mars, and they are generally perceived whenever the telescope is directed upon that planet.

Flammarion went on to suppose that, since the continents of Mars are tinged red, the skies must be red as well:

> In that case the poets of that world would sing the praises of that ardent hue, instead of the tender blue of our skies. In place of diamonds blazing in an azure vault, the stars would be for them golden fires flaming in a field of scarlet; the white clouds suspended in this red sky, and the splendors of sunset, would produce effects not less admirable than those which we behold from our own globe.

And life on Mars seemed almost incontrovertible:

> On earth the smallest drop of water is peopled with myriads of animalcules, and earth and sea are filled with countless species of animals and plants; and it is not easy to conceive how, under similar conditions, another planet should be simply a vast and useless desert.[12]

Taking into account the polar caps, an equatorial zone of continents separated by seas, seasonal changes and the visible fading of surface details at the limb (where the 'ruddy and the greenish tracts are lost in a misty whiteness'), the English astronomer Richard A. Proctor (1837–1888) concluded that Mars is a 'miniature Earth', exhibiting 'clearly and unmistakably the signs which mark a planet as the abode of life'.[13]

While Proctor realized that visual observations were circumstantial and subject to interpretation, what convinced him were the 1867 observations of the English astronomer William

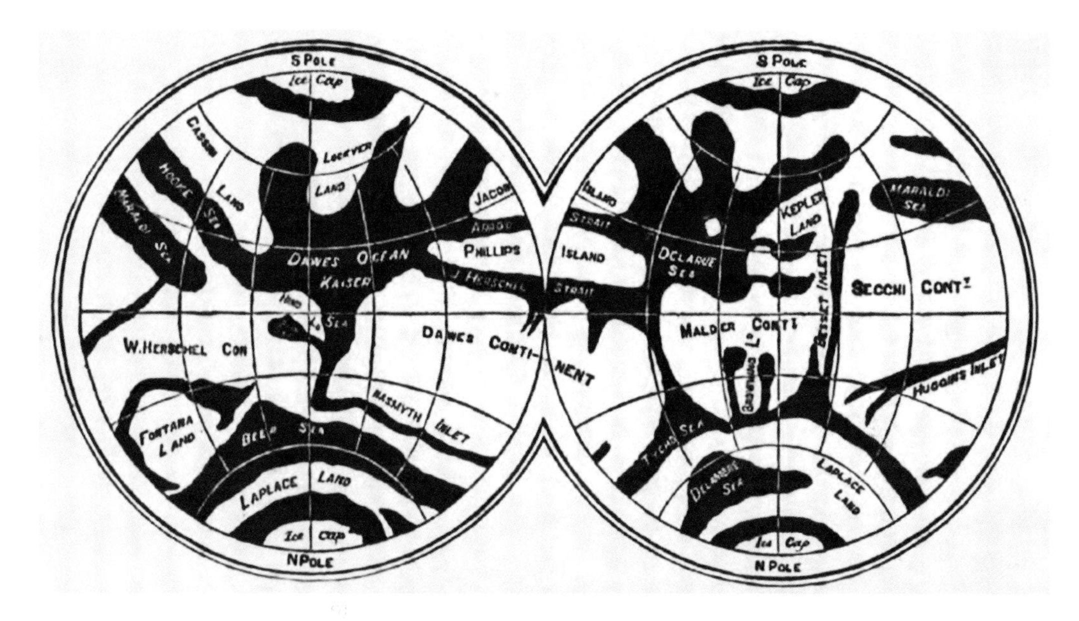

A chart of Mars in stereographic projection by the English astronomer Richard A. Proctor. The map shows the distribution of land masses and seas. The features are named after great Mars observers – a scheme no longer employed.

Huggins (1824–1910), who used a visual spectrograph attached to his 20-centimetre (8 in.) refractor to provide empirical data. The spectrograph focuses incoming light onto a prism or diffraction grating, which separates the light into a spectrum of colours. An object's spectrum displays each chemical element as a unique set of bright (emission) or dark (absorption) lines at specific locations (wavelengths) along the spectrum. As no two elements emit the same spectral lines, we can learn the composition of objects (or parts thereof) by investigating their line spectra.

In the case of objects like the Moon or Mars, which shine by reflected sunlight, the spectrum will be (at first glance) merely a mirror of the Sun's spectrum. And this is what Huggins found in his first attempt during the favourable 1862 opposition of Mars. Five years later, however, he confirmed not only the solar spectrum reflecting off the face of Mars, but a dark telluric line, a band that appears in the spectrum of Earth's atmosphere due to absorption by gases (such as oxygen, water vapour or carbon dioxide). Discovering one in the spectrum of Mars, Huggins said, 'may be accepted as an indication of absorption by the planet, and probably by the atmosphere which surrounds it'.[14]

Huggins found the clouds of Mars especially prominent at the planet's limb and considered they were likely the cause of the limb brightening phenomena, perhaps first noted by Fontana. Due to the obscuring nature of these clouds, as Martian surface features near the limb, they gradually dim. On the other hand, the clouds themselves scatter sunlight, causing the limb to appear bright.

Enter the Canals of Mars

As the analogy to the Earth became ever more compelling, the stage was set for the favourable opposition of 1877. It would be a landmark year in the study of the Red Planet, ushering in an extraordinary episode in the history of telescopic observing. It began when the Italian astronomer Giovanni Virginio Schiaparelli (1835–1910), armed with a 22-centimetre (8.5 in.) refractor at Brera Observatory in Milan, set out to produce a map of the Martian surface of unprecedented accuracy and based on careful measurements of the positions of the various features. In the process, he also devised the innovative system of nomenclature based on classical literature that we still use today.

Earlier maps had given to Martian features the names of astronomers. For example, on Proctor's map there is a Cassini Land, a Mädler Continent and a Beer Sea, while no less than six features bore the name of William R. Dawes, a noted English observer who had made an exquisite series of observations of Mars in 1864. On the whole, however, Proctor's map was chauvinistic as well as redundant – assigning the names of English astronomers, sometimes repetitively, to most of the features – and was immediately doomed on the Continent.

Schiaparelli's system, by contrast, drew on the mythology and geography of the ancient world. On his map the name Solis

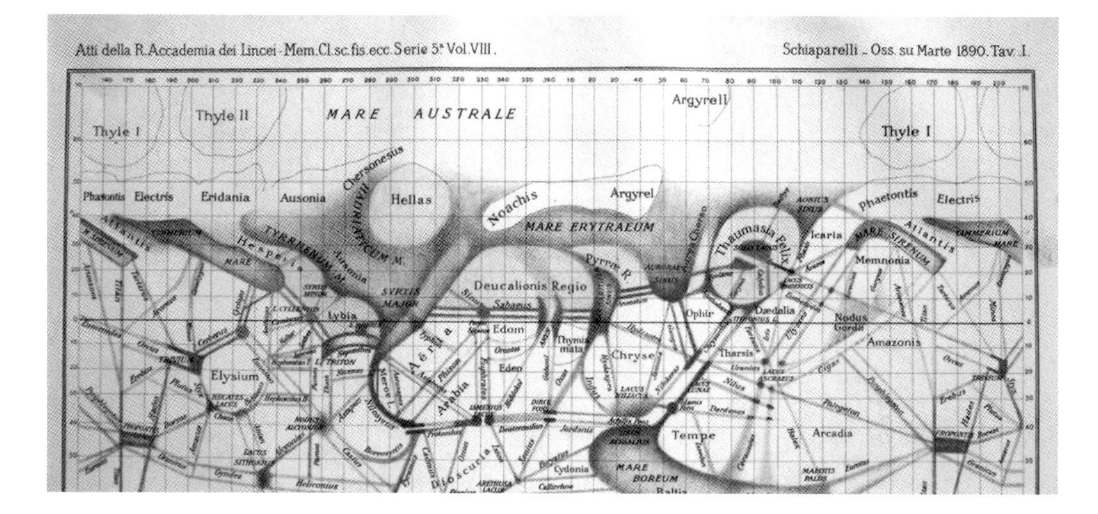

Giovanni Schiaparelli's 1890 map of Mars, with classical names for surface features, canals and much evidence of 'gemination', referring to the doubling of canals.

Lacus, the 'Lake of the Sun', was given to the prominent round patch that had sometimes been called Oculus or the 'Eye of Mars'; it recalled Homer's Bath of the Ocean, from which the Sun rose each morning. To the east was Aurorae Sinus, the Bay of Dawn, in turn followed by Margaritifer Sinus (the rich pearl-bearing Gulf of the Indian Coast), Syrtis Major (the Gulf of Sidra) and Mare Tyrrhenum (the Tyrrhenian Sea, west of Italy). The bright areas included Hellas, Arabia and Elysium. The poetry of these names was undeniable and they soon came into general acceptance.

In addition to charting the main dark and light patches on the planet, which Schiaparelli and other astronomers of the time regarded as the seas and continents of Mars, the Italian astronomer glimpsed a network of fine linear features on the surface, to which he gave the name *canali*, a word first used by Secchi to describe the Hourglass Sea feature now known as Syrtis Major. The popular press translated *canali* into 'canals', with the predictable result that during the 1890s it became all the rage for observers to hunt down and see the 'canals'. Schiaparelli's later maps showed the 'canals' in ever more regular and geometric form. He also described a strange phenomenon that he called 'gemination': sometimes a single 'canal' would appear in a given course, at other times there would be, inexplicably, two running side by side in parallel.

At first Schiaparelli firmly maintained that the 'canals' were natural surface features: even as late as 1893 he was still writing, 'The network formed by them was probably determined in its origin in the geological state of the planet . . . It is not necessary to suppose them the work of intelligent beings.' Two years later, however, still puzzled by the geminations, he suggested that the idea that intelligent beings were behind them 'ought not to be regarded as an absurdity'.[15]

Schiaparelli's thinking may have been influenced by the monumental fervour for building artificial canals on Earth,

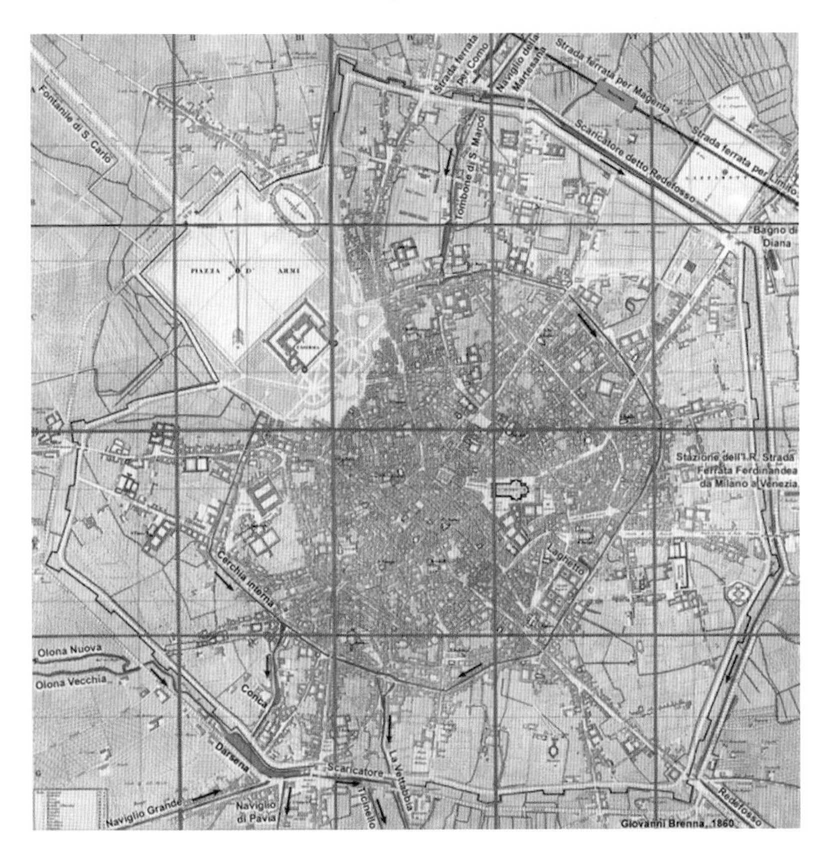

City plan of Milan, Italy, 1860, showing the major navigli canals at their best.

including the 1869 completion of the Suez Canal after the Suez plains reservoir was breached and waters of the Mediterranean flowed into the Red Sea, and the ongoing efforts to construct a canal across an 80-kilometre (50 mi.) stretch of the Panama isthmus. At Brera, Schiaparelli even observed Mars from the banks of the navigli of Milan, a geometrical network of interconnected artificial canals that had been one of the largest medieval engineering projects, based on original ideas from Leonardo da Vinci; these canals surrounded and sliced through the city, connecting it to some of the region's lakes and rivers, and ultimately to the Adriatic Sea. Indeed, during the infancy of canal hysteria, observers could have been describing the navigli when they saw the straight lines or canals on Mars forming a geometrical network extending all over the planet's continents.

Canal fever spilled over to the shores of North America, where Percival Lowell (1855–1916), a Boston entrepreneur who had become a millionaire by the age of thirty by managing his grandfather's textile business, had developed an all-consuming passion for the Red Planet by 1893. His imagination captivated by Schiaparelli's discoveries, Lowell decided to carry out his own study of Mars in Flagstaff, Arizona. Constructing an observatory, equipped with an 11-centimetre (4 in.) Clark refractor and a 46-centimetre (18 in.) Brashear refractor on loan from Harvard, Lowell and his two Harvard assistants, William H. Pickering and Andrew Douglass, opened the Observatory's shutters at the end of May 1894, in time to observe the last favourable Martian opposition of the century, due in October 1894. By July 1896 Lowell had replaced the borrowed telescopes with one he purchased for $20,000, a 58-centimetre (23 in.) refractor from the firm of Alvan Clark & Sons, which had one of the finest optics of that size ever built.

Lowell's observing notebooks reveal that he came to Flagstaff sharing, understandably, the Schiaparellian belief that the dark areas on Mars were seas; after all, the dark areas often appeared sea-blue or green. However, when Pickering trained an instrument called a polariscope on Mars, he unexpectedly found that the light reflected from the dark areas was not polarized. As light reflected from water is polarized, there could be only one conclusion: the bluish-green areas were not seas. But if not, what were they?

Definite changes seemed to have taken place in the outline of the seas over time, which earlier astronomers had regarded as evidence of watery inundations or retreats. In his 1895 book *Mars*, Lowell described the changes he had observed from Flagstaff: 'The long, dark streaks that in June had joined the Syrtis Major to the polar sea had by October nearly disappeared; in their southern parts they had vanished completely, and they had very much faded in their northern ones.'[16] Furthermore, the extent of these changes would prove to be enormous, including a wholesale transformation of the blue-green regions into orange-ochre ones. If there were no seas on Mars, how could the dark features be changing?

As the observed changes on Mars spanned only Martian spring to late summer in the planet's southern hemisphere, Lowell could think of only one theory to explain them: the bluish-green areas were tracts of vegetation, whose changes of colour from greenish to yellow corresponded to the leaf-colour changes we see during autumn on Earth in the northern hemisphere. As for the large ochre areas of the planet, Lowell imagined them to be vast stretches of arid deserts, reminiscent of the badlands that painted the desert south of his observatory. Coincidental or not, Lowell's Martian deserts (as inspired by his desert landscape) and Schiaparelli's imagining of canals (from his canal-infused surroundings) melded into a fantastic idea.

The planet's smaller size, Lowell argued, had caused it to evolve more rapidly than the Earth; it had cooled more quickly and had consequently lost most of its water and air. The only significant sources of water were the snowcaps at the poles, whose thinness was proved by the rapidity of their melting. As the caps melted, they were surrounded each spring by a narrow blue band hugging their retreating edges, a band Lowell identified as meltwater. Given the dearth of water on Mars, Lowell suggested that if beings of sufficient intelligence inhabited it, they would probably be forced to resort to a system of irrigation to pump water from the poles to the desert regions – an engineering project easier to carry out on Mars than on Earth because of the supposed flatness of the planet. It would, in fact, presumably look in plan precisely like Schiaparelli's map.

It is rather odd, as David Sutton Dolan points out, that Lowell's Martians, despite their supposedly advanced intellects, 'were assumed to have no more technically advanced (as opposed to larger-scale) means for dealing with problems of food production and supply. In its reliance on canals, earthworks and agriculture, Lowell's idea of Martian technology was firmly based on the culture of his own terrestrial era as always.'[17]

Indeed, not all astronomers accepted Lowell's views; in fact, most did not. The debate over the 'canals' reached a climax during the close opposition of 1909. Eugène M. Antoniadi (1870–1944), Flammarion's former assistant at his Juvisy Observatory, was granted permission to use the 0.8-metre (2.6 ft) refractor at Meudon Observatory, near Paris, which was then and still is the largest such instrument in Europe. On 20 September 1909, his first night of observing Mars with it, Antoniadi enjoyed several hours of 'glorious seeing'. In fact, his views that night would prove to be the

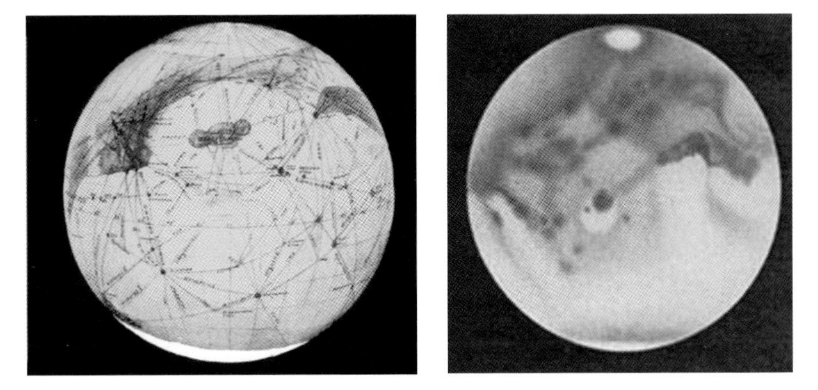

Percival Lowell's globe of Mars and its canals, 1905, and (right) the similar face of Mars by Eugène Michael Antoniadi, 1909.

best of a long career; the Martian deserts were resolved before his eyes into 'winding, knotted, irregular bands, jagged edges of half-tones', or other complex details – almost everything, that is, *except* canals.[18]

In his 1930 book *The Planet Mars*, Antoniadi includes a chapter titled 'The Illusion of the Canals', in which he notes how the British astronomer E. Walter Maunder (1851–1928), 'to whom science was the most important contributor to the theory of the alleged canals', interpreted them simply as 'the sum of a complexity of details'.[19] The Italian astronomer Vincenzo Cerulli (1859–1927) supported this idea, saying that they were optical illusions, whereby a series of dark features run together under imperfect atmospheric conditions to form a line, as was Antoniadi's own experience.

Antoniadi clarifies that he does not doubt Schiaparelli's observations, only his interpretations of them; Schiaparelli's double canals at least have a basis in reality, he said, being real features seen as double lines under eye fatigue, which Antoniadi claims to have witnessed himself. On the other hand, the canals drawn by Lowell and his assistants are 'completely illusory', adding that several authors have compared Lowell's charts of Mars sarcastically with spiderwebs.

Antoniadi's hostility towards the geometry of Mars did not provide the *coup de grâce* to the 'canals', but it certainly weakened the case for them. Nor did his arguments sway Lowell, who continued to popularize his views in two more books, *Mars and its Canals* (1906) and *Mars as the Abode of Life* (1908), as well as a host of newspaper articles, mass-circulation magazines and eagerly attended public lectures; he remained firmly committed to the existence of canals, as well as an irritant to and the envy of other astronomers, right until his death in 1916.[20]

Since Galileo first turned his telescope towards Mars in 1609, we have seen how the planet evolved in the course of three centuries from a diminutive orb with ill-defined patches, to an Earth-like world replete with its own geography and conditions capable of harbouring intelligent life. Percival Lowell marched at the forefront of the crusade for Mars as an abode of life, conducting a blitzkrieg publicity campaign and sending out for publication a series of articles to promote his findings and conclusions about Mars and its race of dying inhabitants. Never before had the astronomical world seen anything like it. Despite the growing gap between Lowell and his peers, Lowell's book on Mars had served its purpose by infiltrating the public's imagination, setting the stage for what was to become a new romance with Mars that would persist well into the twentieth century.

THREE
Romancing Mars

At the turn of the twentieth century, when Lowell's Mars fervour was still at fever pitch, a 'parallel explosion' of interest in telepathy, reincarnation and clairvoyance took place. As Robert Crossley explains in his 2008 article 'Mars and the Paranormal', during this time Lowell's ability to persuade vast segments of the public to his views on Mars gave the spiritualists hope that they would find an ally in him; Lowell, however, made it clear that he had 'little patience with table rappers and clairvoyants who sought to find in him a kindred spirit'.[1] But while Lowell's campaign for the existence of Martian canals gained popular support, it came under increasing scientific attacks, with the American geologist Eliot Blackwelder denouncing the theory as a 'kind of pseudo-science' that Lowell 'foisted upon a trusting public'. Adding fuel to this fire, some mediums had begun publishing their own imaginings of Mars.[2]

One of the most hallowed accounts came from the Swiss medium Hélène Smith, a pseudonym for Catherine Müller (1861–1929), who had visions of Mars while under hypnosis induced by the eminent psychologist Théodore Flournoy (1854–1920) at the University of Geneva. Smith claimed to have contact with people living on Mars and to be able to speak their language, which conveniently sounded like French and which she wrote down during her seances. She also made sketches of the Martian landscapes she witnessed.[3]

Fig. 12. House of Astané. Blue sky; soil, mountains, and walls of a red color. The two plants, with twisted trunks, have purple leaves; the others have long green lower leaves and small purple higher leaves. The frame-work of the doors, windows, and decorations are in the shape of trumpets, and are of a brownish-red color. White glass (?) and curtains or shades of a turquoise-blue. The railings of the roof are yellow, with blue tips.

Fig 14 Martian landscape. Sky of yellow; green lake; gray shores bordered by a brown fence; bell-towers on the shore, in yellow-brown tones, with corners and pinnacles ornamented with pink and blue balls; hill of red rocks, with vegetation of a rather dark green interspersed with rose, purple, and white spots (flowers); buildings at the base constructed of brick-red lattice-work; edges and corners terminating in brown-red trumpets; immense white window-panes, with turquoise-blue curtains; roofs furnished with yellow-brown bell-turrets, brick-red battlements, or with green and red plants (like those of Astané's house, Fig. 12). Persons with large white head-dresses and red or brown robes.

THE MARTIAN CYCLE AND LANGUAGE

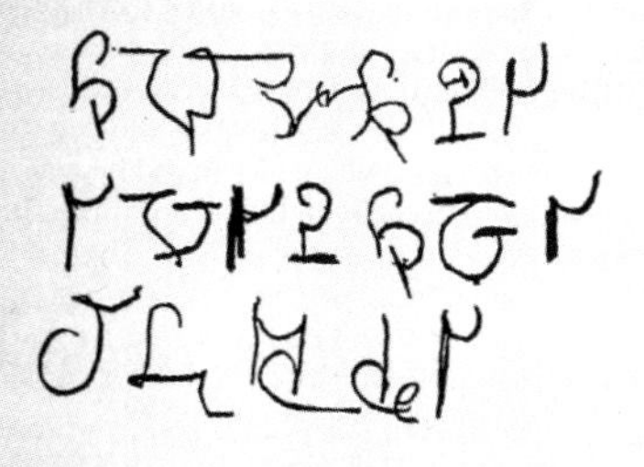

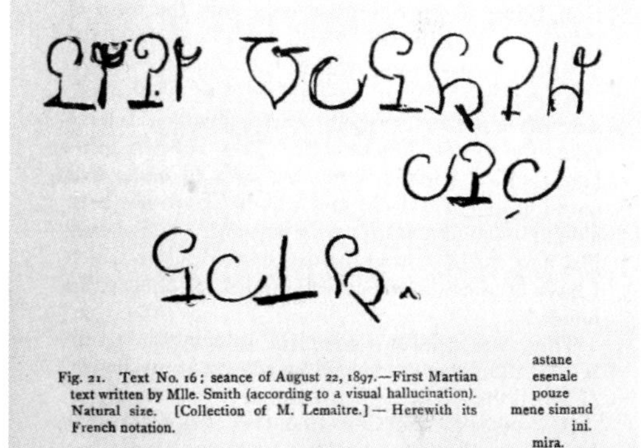

Fig. 21. Text No. 16; seance of August 22, 1897.—First Martian text written by Mlle. Smith (according to a visual hallucination). Natural size. [Collection of M. Lemaître.] — Herewith its French notation.

astane
esenale
pouze
mene simand
ini.
mira.

205

Left: Martian landscapes drawn by Swiss medium Hélène Smith, who received spiritual messages through visual and auditory means as well as raps on the table from her spirit guide and protector named Leopold.

Right: The idea of a special handwriting belonging to the planet Mars, as it occurred, to Hélène Smith's astonishment, during a Martian semi-trance.

The public's mood was more widely grasped with perfect pitch by a thirty-year-old British schoolteacher-turned-writer, Herbert George Wells (1866–1946), who decided to write a book about the Martians, but give them a menacing twist: instead of the wise and peaceful folk that Lowell envisaged, Wells created Martians far more human and believable in his science fiction novel *The War of the Worlds*. We see them glancing across space from their dying world, setting their faces like flint towards ours:

> No one would have believed in the last years of the nineteenth century that this world was being watched keenly and closely by intelligences greater than man's and yet as mortal as his own

> ... No one gave a thought to the older worlds of space as sources of human danger ... Yet across the gulf of space, minds that are to our minds as ours are to those of the beasts that perish, intellects vast and cool and unsympathetic, regarded this earth with envious eyes, and slowly and surely drew their plans against us.[4]

Wells's book, serialized in 1897 and published between covers in 1898, was the first of a voluminous body of works of science fiction, according to cultural historian Robert Markley, 'concerned with depicting the social, political, and economic consequences on Lowell's dying planet'.[5] Other authors soon followed suit: chief among them was Edgar Rice Burroughs (1875–1950), whose *A Princess of Mars* (serialized in 1912) appeared two years after the height of the Mars furore. In it a Virginian soldier, John Carter, could will himself to Barsoom (Mars), where he would roam the dead sea bottoms and subterranean chambers of the dying world – facing death time and again with a fierce Martian civilization in a perilous land, and where science battled against savagery. Ironically, Burroughs's work had appeared when the scientific community was pitted against Lowell and his Martian canal theories, and when Lowell 'frequently fulminated against science, and in fact all mankind, for failing to recognize and acknowledge the cosmic importance of his Mars work'.[6]

Then, on Halloween evening, 1938, the Martians landed in the minds of millions of panic-stricken Americans. The invasion began promptly at 8 p.m. Eastern Standard Time, when Orson Welles and the cast of the Mercury Theatre on the Air took their places before a microphone at a New York studio of the Columbia Broadcasting System and began relaying Howard Koch's freely adapted version of H. G. Wells's *The War of the Worlds*. Listeners who missed Welles's introduction (announcing that what they were about to hear was only a dramatization) turned on their radios and, according to American

Henrique Alvim Corrêa's pen and ink drawing for a 1906 special edition of H. G. Wells's *The War of the Worlds*.

public opinion analyst Hadley Cantril (1906–1969), became 'panic-stricken' by a broadcast that purported to describe an invasion of Martians that threatened our civilization: 'Probably never before have so many people in all walks of life and in parts of the country become so suddenly and so intensely disturbed as they did on this night.'[7] According to Cantril, in his 1940 work *The Invasion from Mars: A Study in the Psychology of Panic*, of the 1.7 million residents who heard the broadcast, 28 per cent believed it was a news bulletin. Seventy per cent of those listeners were frightened or disturbed by it; about 1.2 million people across the nation were affected. Cantril proposed that the unusual realism of the performance attributed to the fact that the early parts of the broadcast fell within the existing standards of judgement of the listeners.[8]

As the incredible one-hour drama unfolded the audience learned that the Martians were hideous beings, having appendages like grey snakes and V-shaped mouths with saliva dripping from their rimless and quivering lips; that they emerged from their cylinders armed with heat rays; that all armed resistance sent against them met with horrifying death; that Martian cylinders were 'falling like flies' all over the country; that the world was near an end. The next day newspapers spoke of the 'tidal wave of terror' that swept the nation. And so, in the public's perception of an inhabited, warlike Mars, expectation triumphed by creating illusion.

These and other works of science fantasy cannot be dismissed as completely irrelevant to the scientific investigation of Mars, as they were an essential ingredient of the romanticism that fuelled the dream of space flight.[9]

A Parallel World

Lowell's views of the canals were based on visual observations with a telescope. The supremacy of the eye for planetary observations – in contrast to stellar and nebular observations, where the greater light-gathering power of the photographic plate quickly gave photography a decisive advantage – continued well into the twentieth century. However, the subjectivity and differences of observers could not be denied, and as early as the 1860s a strong movement was afoot to move away from visual telescopic observations towards those that provided more quantitative information through the use of instruments, such as photometers, spectrometers and thermocouples.

The German astronomers Ludwig Seidel (1821–1896) and Johann Zöllner (1834–1882) were among the early photometric pioneers of Mars. In 1862 they began a study to determine the percentage of sunlight the planet reflects back into outer space (its *albedo*, meaning 'whiteness'; freshly fallen snow, for instance, reflects 95 per cent of the sunlight hitting it, so it has a high albedo of 0.95, with 1.0 being a perfect reflector – freshly laid asphalt, on the other hand, reflects sunlight poorly, having a low albedo of 0.04). They found that the reflective power of Mars is only one-and-a-half times greater than that of the Moon, a low albedo object that reflects only 12 per cent of the Sun's light on average; in contrast, gas giant Jupiter reflects 52 per cent of the Sun's light. Thus the light we see from Mars is almost wholly reflected by the true surface and not cloud.

Spectroscopic observations by William Huggins in the mid-nineteenth century expanded the instrumental investigation to the atmosphere of Mars, when he identified the spectral signature of water vapour (at least in part). Once Huggins focused his spectroscope on the Martian dark areas, however, specifically on the dark areas bordering the south polar spot, he also found a difference in solar absorption compared to the surrounding brighter areas. He deduced that the 'material which forms the darker parts of the planet's surface, absorbs all the rays of the spectrum equally. These portions should be therefore neutral, or nearly so, in colour.' He added the following prophetic note:

> it does not appear to be probable that the ruddy tint which distinguishes Mars has its origin in the planet's atmosphere . . . The evidence we possess at present appears to support the opinion that the planet's distinctive colour has its origin in the material of which some parts of its surface are composed.[10]

Huggins's results were subsequently confirmed by other astronomers, who followed his method and used the spectroscope to find water on Mars – for obvious reasons, since an active water cycle was required to justify the canals and the existence of life. But the picture was imperfect, because, as even the simplest theoretical considerations were able to show, water could not exist upon the surface of Mars without instant evaporation.[11] The more basic question now surrounding the Martian water supply (in both the liquid and vapour states) was how much? The astronomers who continued Huggins's quest included Angelo Secchi in Italy, Hermann Vogel in Germany and, most notably, Pierre Jules Janssen (1824–1907) in France.

French astronomer Pierre Jules Janssen (1824–1907).

Gravitating towards the exciting possibilities of spectroscopy in astronomical research, Janssen made a careful study in the mid-1860s, from high in the Alps, of the absorption of the solar spectrum by the Earth's atmosphere. He concluded that most of the telluric lines superimposed on the solar spectrum were produced by water vapour. He concentrated his attention on these lines in a study of the spectrum of Mars, carried out in May 1867 on the volcano Mount Etna (3,295 metres (10,810 ft) high) in Sicily. Because of the great altitude of Etna and the bitter cold near the summit, very little water existed above his station, and yet he weakly saw the water bands in the spectrum of Mars, though none in that of the Moon, which was then at lower altitude. From his observations he concluded 'that there are powerful reasons for thinking that life is not the exclusive privilege of our small earth, the younger sister of the great family of planets'.[12]

How Inhospitable

As spectroscopic and visual studies progressed, evidence mounted rapidly against earlier assumptions that the dark maria on Mars were seas. One opponent of the interpretations of the dark areas being seas was William Pickering (1858–1938), younger brother of Harvard Observatory's Edward C. Pickering. William's interest in Mars began after the favourable 1877 opposition of Mars, when Asaph Hall at the U.S. Naval Observatory discovered its two moons (see Chapter Eight) and the Italian astronomer Giovanni Schiaparelli announced his discovery of Martian *canali*: 'I am so much interested in Mars,' Pickering wrote to his brother Edward. 'I always thought it the most interesting planet in the system, because its climate so nearly resembles our own. And now it's more interesting than ever.'[13]

Under the direction of his brother, in 1891 William Pickering helped establish the Boyden Station, a high-altitude astronomical observatory in Arequipa, Peru. There, he and his assistant A. E. Douglass observed Mars almost every night from July to September 1892 with a 33-centimetre (13 in.) long-focus refractor of high optical quality. In a series of telegrams to the *New York Herald*, Pickering described canals, mountain ranges near the planet's South Pole, evidence of melted snow and water flowing northward between them, clouds in the Martian atmosphere, and some forty lakes on the surface occurring at the intersections of the canals.[14]

More importantly, he discovered a network of canals cutting across the Martian maria, leading him to doubt that the dark markings, once believed to be seas of water, and the canals carried

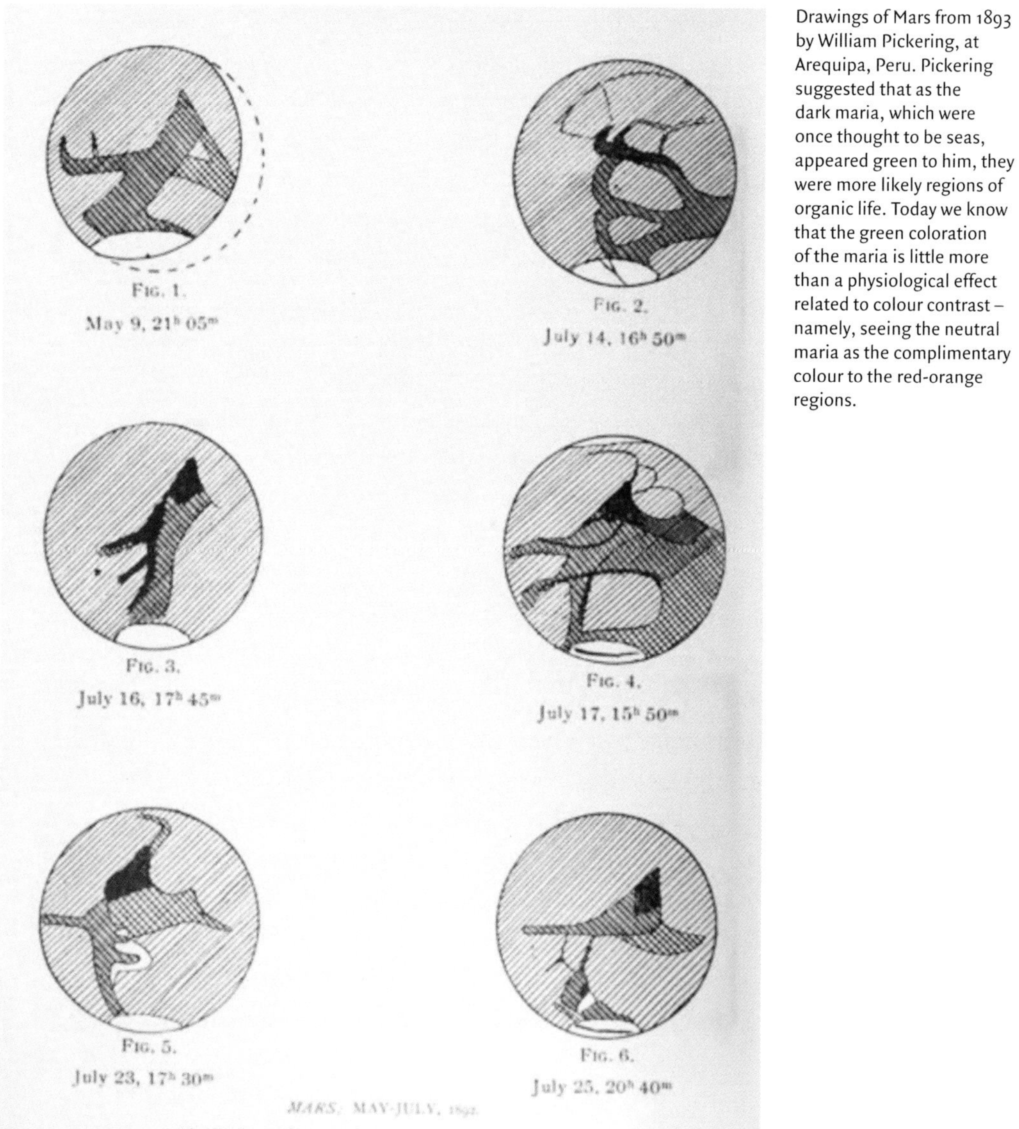

Drawings of Mars from 1893 by William Pickering, at Arequipa, Peru. Pickering suggested that as the dark maria, which were once thought to be seas, appeared green to him, they were more likely regions of organic life. Today we know that the green coloration of the maria is little more than a physiological effect related to colour contrast – namely, seeing the neutral maria as the complimentary colour to the red-orange regions.

any water at all, especially as the mare appeared green. Rather, he deduced, the maria were regions of organic life and the canals were really strips of vegetation growing along the banks of canals that were too fine to be seen (a theory picked by Lowell). By 1905, however, Pickering performed an about-face, modifying his thinking to suggest that the canals were strips of vegetation along the borders of volcanic steam cracks (an image he gleaned from a trip he had taken to the volcanoes on the Big Island of Hawaii). In the end Pickering became critical of Lowell's theories about intelligent life on Mars.

Pickering was not alone. In 1894 William Wallace Campbell turned his attention to the Martian water vapour question. He attached to the 91-centimetre (36 in.) refractor of the Lick Observatory at Mount Hamilton, California, then the world's largest telescope, a new and more powerful visual spectroscope than had ever been used on the Red Planet. Given this advantageous telescope–instrument combination, he expected to find stronger evidence of water vapour in the Martian atmosphere than did his predecessors. The results, however, came as a great surprise. During ten nights in which he attempted to compare the spectrum of Mars with that of the Moon, Campbell failed to detect any strengthening of the water vapour bands whatsoever. This did not, of course, prove that there was no water on Mars, only, according to his calculations, that the amount present must be no more than one-quarter that above Mount Hamilton on a dry summer night.[15] Lowell's dying world suddenly had even less water than could be imagined. He concluded the spectrum of Mars's surface is similar to that of the Moon's, throwing doubt on the idea that the atmosphere of Mars is like Earth's. And the water detected by earlier observers must have been from Earth's atmosphere.

Campbell's results were confirmed in 1925 by Walter S. Adams (1876–1956), who determined from the Lick Observatory that the water vapour in the Martian atmosphere at the time was in the order of 5 per cent of that normally found in Earth's atmosphere, with an oxygen level less than above Mount Everest.[16] Campbell willingly endorsed Adams's work as being in agreement with his own, without any apparent consideration that this work, too, might be illusory – which, indeed, it turned out to be.[17]

Nevertheless, Lowell's book *Mars* was attacked from another flank of science. In 1904 the well-known biologist Alfred Russel Wallace (1823–1913) responded to Lowell's book with one of his own, *Man's Place in the Universe*, in which he argued that Mars is completely uninhabitable:

> Mars receives less than half the amount of sun-heat per unit of surface that we do. And as it is almost certain that it contains no water (its polar snows being caused by carbonic acid or some other heavy gas) it follows that, although it may produce vegetable life of some low kinds, it must be quite unsuited for that of the higher animals.
>
> Its small size and mass, the latter only one-ninth that of the earth, may probably allow it to possess a very rare atmosphere of oxygen and nitrogen, if those gases exist there, and this lack of density would render it unable to retain during the night the very moderate amount of heat it might absorb during the day. This conclusion is supported by its low reflecting power, showing that it has hardly any clouds in its scanty atmosphere.
>
> During the greater part of the twenty-four hours, therefore, its surface-temperature would probably be much below the freezing point of water; and this, taken in conjunction with the total absence of aqueous vapour or liquid water, would add still further to its unsuitability for animal life.[18]

By 1945 astronomers could readily summarize the most rudimentary evidence against the Martian seas: first, the brilliant reflection of the Sun from an open-water source has never been observed; second, the dark markings exhibit a great amount of finer detail that would not be possible in a body of water; and finally, the dark areas show variation in shade and depth of colour, and size and shape, as the Martian seasons progress. This left only the polar caps, and its feeble clouds, frosts and mists, as the only possible sources for water vapour.[19]

The view of Mars as a possible abode of life continued to fade. Beginning in November 1923 the American astronomers Edison Pettit (1889–1962) and Seth Nicholson (1891–1963) used a vacuum thermocouple on the 254-centimetre (100 in.) Hooker telescope on Mount Wilson, California, to determine thermal radiation emitted by Mars's surface. Their observations, conducted at intervals over the course of a year, revealed that Mars is much like Mercury and the Moon, displaying great swings in temperature: during the day, areas of Mars can achieve a temperature just above freezing; the temperature at night can dip to –85°C (–121°F); and the mean temperature of the polar caps is about –70°C (–94°F).

The French astronomer Bernard Lyot (1897–1952) added to the idea that Mars was a Moon-like, desolate planet. All bodies that reflect – including the Moon and Mars – are also polarized to some degree. Armed with that knowledge, Lyot used, between 1922 and 1928, a powerful instrument particularly sensitive to polarized light to analyse the amount coming from Mars compared with that from the Moon. His results showed that the polarized light coming from the two worlds were precisely similar. He furthered his studies in the laboratory, finding that a mixture of grey and brown volcanic ash displayed the same degree of polarization.

Until this time, the nature of the planet's atmosphere had remained elusive. Decades of spectroscopic studies had failed to provide any solid evidence concerning the amounts of gases that might be present relative to Earth's, leading only to suggestions. That situation changed in 1947 when the Dutch-American astronomer Gerard Kuiper (1905–1973) struck yet another blow to the theory that Mars's atmosphere contains a noble amount of water. By simultaneously comparing the infrared spectrum of Mars to that of the Moon, he detected definite evidence that the Martian atmosphere contains an abundance of carbon dioxide, more than is found in Earth's atmosphere; he confirmed this finding a year later when Mars was closer to Earth.

By 1963 additional studies had shown that the abundance of carbon dioxide was even greater than previously thought – about 17 per cent. That same year the French astronomer Audouin Dollfus (1924–2010) took the spectroscopic studies of Mars to a new height – literally to a height of 14,000 metres (46,000 ft; more than four times the height of Mount Etna) – when he ascended in a balloon from the terrace of Meudon Obsevatory in search of the elusive knowledge of how much water vapour invades the atmosphere of Mars. The amount he found was shockingly small: the equivalent of 1 cubic metre of liquid water for the whole planet, whose surface area is equal to that of all of Earth's continents joined together. Dollfus's finding gives us some idea of the terrible reality of the Martian deserts.

At the dawn of the twentieth century, when highly imaginative views of Mars inspired by science-fiction fantasy writers and clairvoyants, who used Lowell's theories on Mars as a foundation for their works, instilled a provocative and at times frightening portrait of Mars as an inhabited world in the public's eye, scientific investigations continued to gather evidence to the contrary. Even so, Earth-based studies of Mars remained difficult, and progress in deciphering the Martian atmospheric code was gradual at best.

Despite this headway, the Lowellian version of Mars – minus canals, perhaps, but still with tracts of vegetation and lonely deserts – still had a pulse. A 1959 U.S. Congressional report on space activities, for instance, which included an update on Mars, stated that 'there is rather good evidence that some indigenous life forms may exist' on the planet, noting specifically that in 1958

Harvard astronomer William M. Sinton (1925–2004) 'has found spectroscopic evidence that organic molecules may be responsible for the Martian dark areas'.[20] Such thinking continued right up until July 1965, when NASA's *Mariner* 4 spacecraft passed some 1,000 kilometres (620 mi.) from the planet and sent back to Earth the first images of its surface, starting a new chapter in the long and oscillating history of humanity's quest to understand this baffling world.

The three men responsible for the success of *Explorer 1*, America's first Earth satellite, launched 31 January 1958. At left is William H. Pickering, former director of JPL, which built and operated the satellite. James A. Van Allen, centre, of the State University of Iowa, designed and built the instrument on *Explorer* that discovered the radiation belts which surround the Earth. At right is Wernher von Braun, leader of the Army's Redstone Arsenal team which built the first stage Redstone rocket that launched *Explorer 1*.

FOUR

THE FIRST EMISSARIES TO MARS

The Soviet Union ushered in the space age with the launch on 4 October 1957 of *Sputnik 1*, humanity's first Earth-orbiting satellite. The launch also triggered a Cold War hysteria in the United States; now that the Soviets had beaten the U.S. into space, were they, then, technologically superior? Did they now have the capability to rain down nuclear warheads onto U.S. soil from space? As George E. Reedy, who served as White House Press Secretary under President Lyndon B. Johnson from 1964 to 1965, later recalled, many Americans at the time felt that 'we can no longer consider the Russians to be behind us in technology. It took them four years to catch up to our atomic bomb and nine months to catch up to our hydrogen bomb. Now we are trying to catch up to their satellite.'[1]

Before the effects of *Sputnik 1* had a chance to simmer, the Soviets followed with a second *Sputnik* spacecraft, which also achieved Earth orbit; both craft were inspired by the International Geophysical Year (IGY), an international scientific project that lasted from 1 July 1957 to 31 December 1958, during which scientists from around the world began an intense year of research on satellites.

The Eisenhower administration moved quickly to 'restore confidence at home and prestige abroad' by announcing a much-hyped U.S. Navy test launch of a Project Vanguard booster on 6 December 1957.[2] During the ignition sequence, the rocket rose about 1.2 metres (4 ft) above the platform, shook briefly and disintegrated in flames. The media event was a disaster, earning the names 'Flopnik' and 'Kaputnik'.[3]

Meanwhile, the U.S. Army Ballistic Missile Agency was directed to launch a satellite using its Jupiter C rocket, developed under the direction of Dr Wernher von Braun (1912–1977). The Jet Propulsion Laboratory (JPL) received the assignment to design, build and operate the artificial satellite that would serve as the rocket's payload. JPL completed this job in less than three months.[4] After two launch aborts, the Juno 1 booster carrying *Explorer 1* lifted off from Cape Canaveral, Florida, on 31 January 1958 and entered Earth orbit, where its 203-centimetre-long (80 in.) satellite made the first scientific discovery in space: Earth is surrounded by radiation belts of electrons and charged particles, some of them moving at nearly the speed of light (*c*. 300,000 km per second).

In the wake of these and further launch initiatives – both successes and failures – events moved quickly towards the development of a permanent federal agency dedicated to exploring space. On 1 October 1958, less than a year after the launch of *Sputnik*, the National Aeronautics and Space Administration (NASA) was born and started functioning.

Meanwhile the Soviets had begun to redirect their gaze from lowly Earth orbit to more lofty and ambitious targets: the Moon and its far side, followed by the Red Planet Mars. Ironically, for at least a decade Mars had become a U.S. cinematic symbol of communism during the Great Red Scare and McCarthy 'witch hunts'. These films, including the iconic *Red Planet Mars* (1952), in which fabricated radio broadcasts from a Utopian Mars save Earth by sending out messages that inspire the overthrow of communism and the removal of Soviet troops from around the world, fed the country's paranoia by using a secret invasion of aliens from the Red Planet as a metaphor for subversive communist activity.

After three more successful *Sputnik* launches in the 1950s, including *Sputnik 2* (with the dog Laika) and *Sputnik 4*, which was the test run of the equipment that would safely send cosmonaut Yuri Gagarin (1934–1968) into space and back to Earth, the Soviets made the first attempt to 'invade' Mars with twin spacecraft (*Marsnik* 1 and 2).[5] Launched four days apart in October 1960 on intended Mars flybys, these craft failed to reach Earth orbit. Their failures, however, were greatly overshadowed by the much more astonishing Soviet achievement of sending the first human into space, when, on 12 April 1961, cosmonaut Yuri Gagarin blasted into Earth orbit aboard the Soviet *Vostok* 1 spacecraft.

Still the Soviet push for Mars remained strong. On 24 October 1962 the Soviet *Sputnik* 22 spacecraft achieved Earth orbit, but either broke up or exploded during the burn to put the spacecraft into Mars trajectory.[6] Success followed only eight days later, when the Soviet *Mars* 1 spacecraft became the first to escape Earth's gravity and sail past Mars; although communications had been lost before the craft reached the planet, we know its trajectory carried it within 193,000 kilometres (120,000 mi.) of the planet's surface on 19 June 1963, after which the spacecraft entered an orbit around the Sun.[7]

When the United States entered the race to Mars in 1964, its first attempt ended in failure: shortly after launch on 5 November, NASA's *Mariner 3* spacecraft failed to eject a protective shield and the added weight prevented the craft from attaining its prescribed Mars trajectory.[8] The Soviets made another flyby attempt with *Zond-2*. Launched on 30 November 1964, it flew by Mars unnoticed on 6 August 1965 after a series of mishaps, including a communications failure en route to Mars led to no further contact with the craft.

History was finally made on 15 July 1965 when NASA's *Mariner* 4, after 228 days of interplanetary travel, passed some 1,000 kilometres (620 mi.) from the surface of Mars and snapped the first images of another planet from space. A new era of planetary studies from space had begun.

Data transmission from the *Mariner* 4 spacecraft was painfully slow; the spacecraft could transmit information at a maximum rate

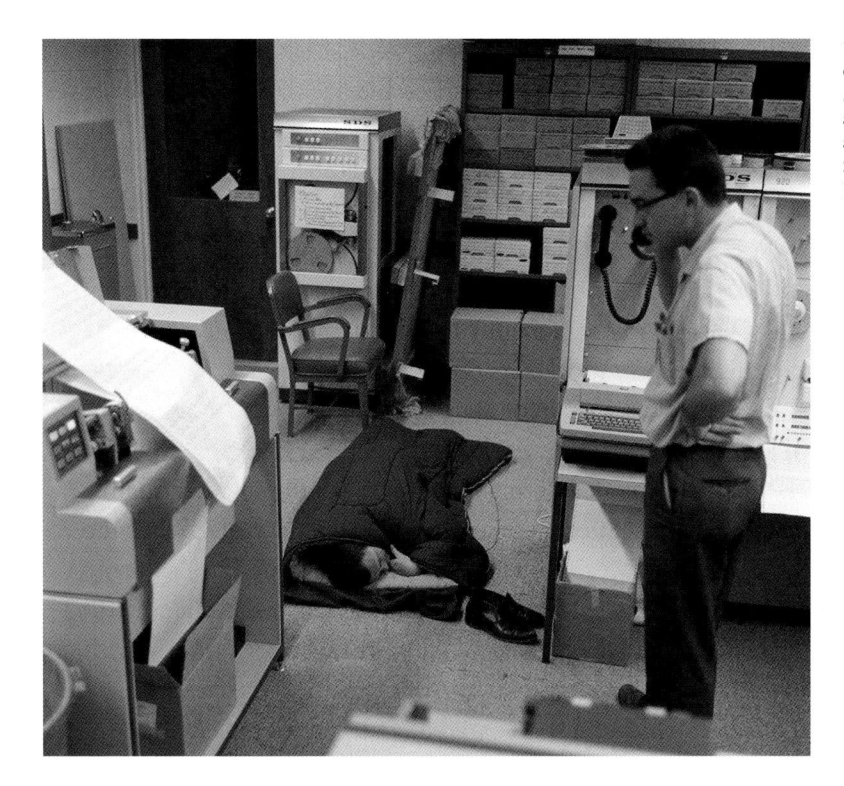

NASA scientists await data returning from *Mariner 4* in the 'Echo' antenna receiving room at Goldstone Tracking Station in California's Mojave Desert.

of only about 8 bits per second, requiring ten hours to send one 200 × 200 pixel image.[9] By comparison, an average smartphone today can download a typical 24-megabyte image in one second on a 4G mobile network. Although the images were blurry, their resolution was equal to the best telescopic views of the Moon from Earth.

While the idea of Martian canals had been all but discredited by the time of *Mariner 4*, the spacecraft revealed no clear evidence of them, though some still held out hope, based partly on William Sinton's spectroscopic results, for the existence of some forms of vegetation, such as lichens. But the most arresting single phenomenon of the entire mission was the discovery of densely packed lunar-style impact craters on the Martian surface, which 'amazed the watching group of scientists'.[10] In the brief 25 minutes of the flyby, the spacecraft's 21 images had crushed centuries of romantic visions about the Red Planet as an abode of life; here instead was a pockmarked world as barren and desolate as the Moon. Given that the first image showed more than seventy craters, this indicated that Mars's surface might be covered by about 10,000.

The first official statement on *Mariner 4*'s results came to many as a devastating blow: at least part of the surface of Mars was heavily cratered and ancient, perhaps 2 to 5 billion years old. In one swipe of a snapping camera, Martian canals and the super-intelligent race that built them vanished from view. And while the spacecraft was too far away to see life in the form of plants or animals, a *New York Times* editorial on *Mariner 4* summarized the results: 'Its surface bathed in deadly radiation from outer space, [Mars] has very little atmosphere and has probably never had large bodies of water such as those in which life developed on this planet.' Only a few 'diehards', the editorial continued, 'refused to give up the ghost and accept Mars as a barren world'.[11] To many, Mars was dead.

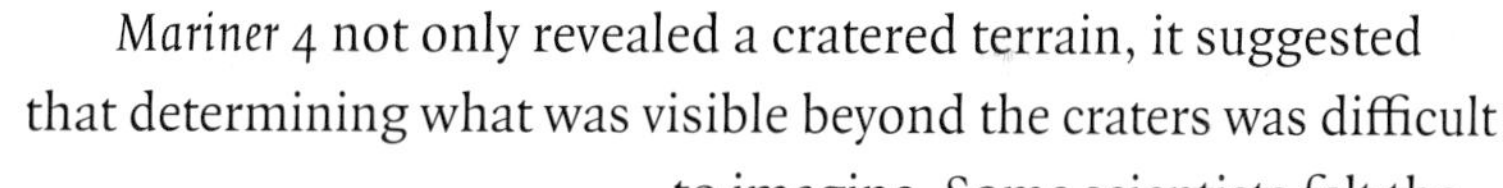

Mariner 4 not only revealed a cratered terrain, it suggested that determining what was visible beyond the craters was difficult to imagine. Some scientists felt the initial reactions to the images were premature, as they lacked the clarity needed to see Mars's more Earth-like features. Had *Mariner 4* imaged Earth at a similar resolution, it would have missed any signs of life. Besides, the images missed 99 per cent of the planet's surface.

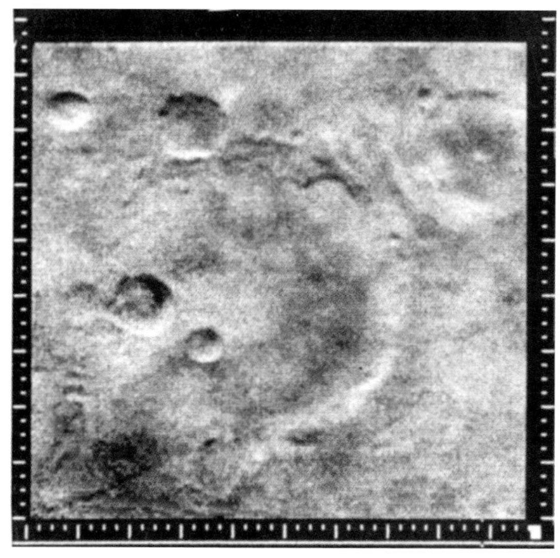

Mariner 4 snapshot of the surface of Mars revealing a lunar-like landscape.

With so much unknown, the exploration of Mars continued. Between February 1969 and May 1971 NASA launched four and the Soviets five spacecraft to Mars. Several of these failed. The first two sent by the U.S. (*Mariners* 6 and 7) succeeded, similarly passing over heavily cratered regions of Mars and producing a few images that served to enhance the bleak, albeit limited, view of the planet. To gain clarity, on 30 May 1971 NASA launched its first long-term mission to Mars, *Mariner* 9, which beat two Soviet spacecraft (*Mars* 2 and 3) to the Red Planet to become the first spacecraft to orbit another world.[12] *Mars* 2 released a descent module four-and-a-half hours before reaching Mars on 27 November 1971, but the module entered the Martian atmosphere at a steeper angle than planned and crashed, nevertheless delivering the Soviet Union coat of arms to the surface.

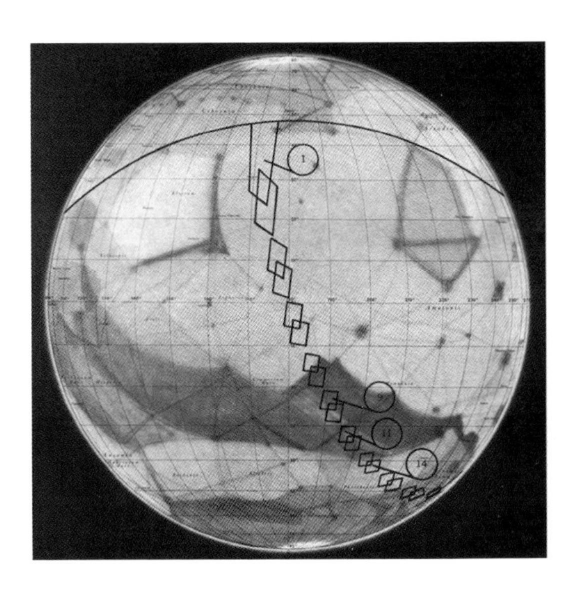

The location map used by NASA to plan the *Mariner* 4 encounter included canals. The trail of 'footprints' show the regions covered by the spacecraft's camera on its 14–15 July 1965 flyby.

During its mission, which lasted nearly a year, *Mariner* 9 succeeded in mapping 85 per cent of the Martian surface. This task, however, was delayed by several months by a dust storm, which started on 22 September 1971 in the Noachis region of the planet and quickly grew into the greatest dust storm observed in a century.[13] The mission's collection of more than 7,000 images rewrote the book on Mars by discovering channels, polar-layered materials, enormous shield volcanoes, sand dunes and a vast canyon system.[14]

The Soviet *Mars* 2 and 3 spacecraft, which followed *Mariner* 9 into Mars orbit by only a matter of two weeks, also took a total of sixty images that revealed the Martian volcanoes and other surface features, enabling the creation of some of the first surface relief maps of the planet. Data transmitted to Earth also revealed the presence of atomic hydrogen and oxygen in the planet's upper atmosphere, as well as water vapour concentrations some 5,000 times less than in Earth's atmosphere, surface temperatures ranging from −110°C (−166°F) to +13°C (55°F), and dust grains as high as 7 kilometres (4 mi.) in the atmosphere.[15] What's more, *Mars 3* released a descent vehicle that parachuted into fame, becoming the first probe to make a soft landing on Mars; it transmitted data for twenty seconds from the surface before it stopped, possibly due to a dust storm.

The *Mariner* and *Mars* spacecraft revealed the Red Planet as never before: a mysterious new world with astonishing features. Our perception of Mars oscillated once again, changing from a frigid, crater-pocked world to a former geological wonder on whose surface ancient riverbeds provide evidence that water may have once flowed across its surface in the distant past. 'If we accept the possibility that life arose on the planet during this earlier epoch,' stated California Institute of Technology biologist Norman Horowitz (1915–2005), 'then we cannot exclude the possibility that Martian life succeeded in adapting to the changing conditions and remains there still today.'[16]

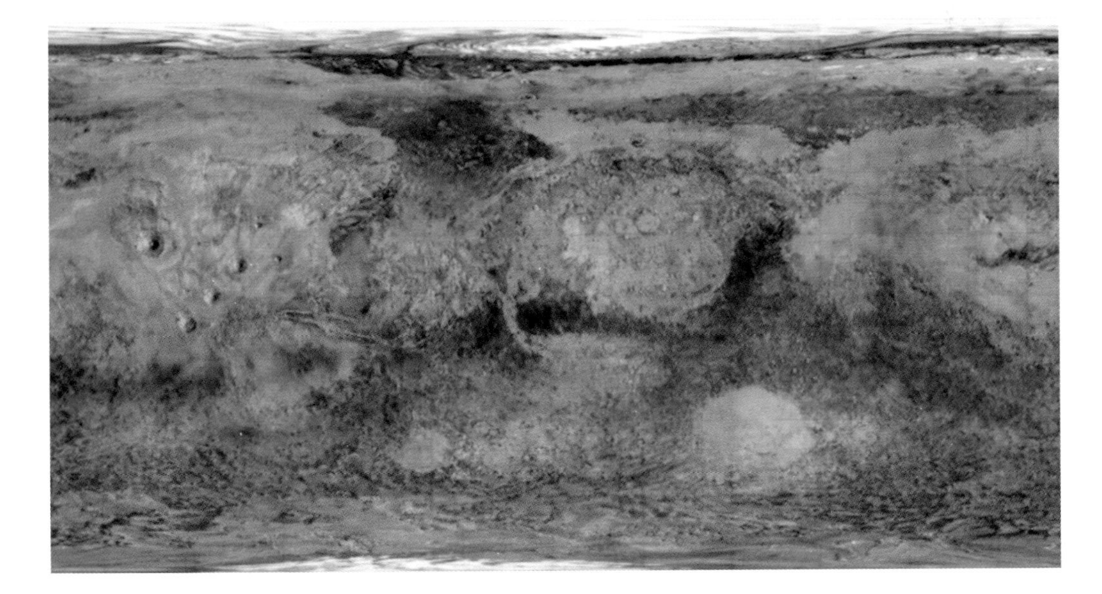

Mariner 9 was the first Mars orbiter. The more than 7,000 images it took provided the first detailed views of all of Mars, revealing that the planet had both ancient cratered terrain and more modern tectonized and eroded areas.

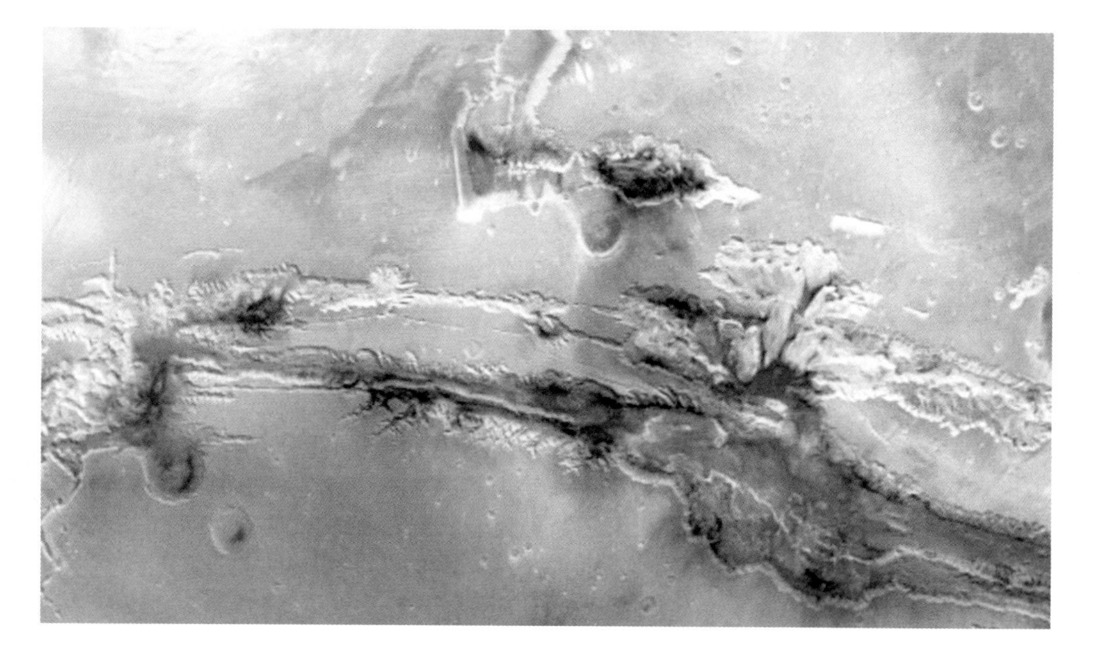

The American astronomer and astrobiologist Carl Sagan (1934–1996) supported the view of possible life on Mars, especially in microenvironments (certain specialized zones of the planet). Sagan speculated that if microbial life were to exist, its size would need to be large enough to protect it against the planet's cold, dry environment and intense ultraviolet radiation. In the late 1960s some researchers were considering terrestrial analogues, such as the kangaroo rat, which can survive for long periods without liquid water.[17]

Images from *Mariner 9* showing the great canyon Valles Marineris and (below) the tallest volcano in the solar system, Olympus Mons.

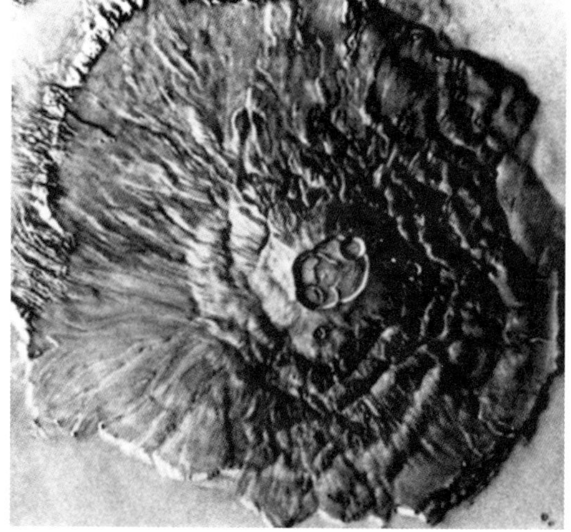

In 1975, after the fresh views of Mars produced by the *Mariner* and *Mars* spacecraft sparked a flurry of new NASA missions, Sagan announced that we are 'only now beginning an adequate

On 10 March 2013 NASA's *Mars Reconnaissance Orbiter* (MRO) took this image of what may be the *Mars 3* lander's parachute spread out to nearly its full extent (11 m) on the surface of Mars.

Taken by the *Viking 1* lander shortly after it touched down on Mars, 20 July 1976, this image is the first photograph ever taken from the surface of Mars. It shows the footpad because scientists and engineers wanted to see how far Viking might sink into the surface. One-third of the scientists thought the consistency of Mars might be like that of shaving cream. It wasn't.

reconnaissance of our neighboring world. There is no question that astonishments and delights await us.'[18]

The Soviet Union continued its *Mars* series of spacecraft (4–7), but only met with partial success: in 1974 one orbiter (*Mars* 5) collected data for 22 orbits, and one lander (*Mars* 6) returned the first data from Mars's atmosphere for 224 seconds before transmission ceased. NASA triumphed again, however, in the summer of 1976 with the arrival of its twin *Viking* orbiters (*Viking* 1 and *Viking* 2) and their respective landers, thanks in part to *Mariner* 9's confirmation of the planet's atmospheric pressure, which allowed engineers to design the *Viking* landers for a safe descent.

The primary *Viking* mission objectives were to obtain high-resolution images of the Martian surface, characterize the structure and composition of the atmosphere and surface, and search for evidence of life. The orbiters studied the planet for more than two Martian years, imaging its entire surface at a resolution of 150 to 300 metres (490 to 985 ft). They imaged a vast area of stunning and sometimes baffling terrain. The *Viking* orbiter views of Mars revealed a mysterious world that could be divided into two disparate regions: a lightly cratered northern lowlands (a vast and arid plain) and highly cratered southern highlands. Superimposed on these are some of the solar system's most fascinating geological wonders, including two enormous volcanic bulges – Tharsis (more than 8,000 kilometres (4,970 mi.) across and 8 kilometres (5 mi.) high, and including Olympus Mons, which rises to nearly 22 kilometres (14 mi.) and Elysium (above which Elysium Mons rises to 13 kilometres (8 mi.)) – and Valles Marineris, a system of giant canyons near the equator that measures more than 3,000 kilometres (1,865 mi.) long, as much as 600 kilometres (375 mi.) across, and as much as 8 kilometres (5 mi.) deep.

The *Viking* 1 lander transmitted the first images of the planet's surface. The very first image showed the craft's footpad because, according the Planning Director B. Gentry Lee, 'One-third of the

Left: *Viking* 1 lander image of the surface of Mars taken on 21 July 1976 – the day after it successfully landed on the planet.

Right: First colour image of the *Viking* 2 landing site. The colours of the rocks and soil are almost identical to that seen by *Viking* 1 more than 6,400 km away. Notice how the sky is pink at both landing sites from airborne dust.

scientists thought the consistency of Mars might be like that of shaving cream.'[19] The *Viking* landers, on the other hand, transmitted images of the surface. The *Viking* 1 site in Chryse Planitia showed a low plain covered with large rocks, loose sand and orange-red dust: a veneer of possible hydrated ferric oxide. Such weathering products form on Earth in the presence of water and an oxidizing atmosphere. *Viking* 2's landing site in Utopia Planitia was much rockier, suggesting the rocks were either fragments of an ancient lava flow, or perhaps ejecta from a nearby crater.

The science instruments aboard the landers acquired more data than expected, revealing that surface material at both landing sites can best be characterized as a type of iron-rich clay that contains a highly oxidizing substance that releases oxygen when it is wetted. Researchers observed only minor changes in the appearance of the near surface over the mission's lifetime. Temperatures at the landing sites ranged from 150 to 250°K, with a variation over a given day of 35 to 50°K. The weather instruments recorded Martian winds generally blowing more slowly than expected; scientists had expected them to reach speeds of several hundred km/h from observing global dust storms from Earth, but neither lander recorded gusts over 120 km/h, and average velocities were considerably lower.

The landers also found cyclic variations in Mars's weather patterns, possibly due to alternating cyclones and anticyclones. The spacecraft recorded nitrogen as a significant component of the Martian atmosphere, an element that had eluded detection until then.[20]

Eleven days after *Viking* touched down, Sagan found himself confronted by a press thirsty for information about whether life existed on Mars. Sagan, who had given himself the task of scrutinizing the first incoming images for signs of movement on the Martian surface, could only reply, 'So far, no rock has obviously got up and moved away.' Sagan and others in the *Viking* exobiology team, who were hoping to see some signs of life, were disappointed with the *Viking 1* lander view: 'There was not a hint of life – no bushes, no trees, no cactus, no giraffes, antelopes, or rabbits,' Sagan quipped.[21] When the second *Viking* lander touched down, it too revealed only a forest of rocks, offering no hope of large life forms. While these sites were geological wonderlands, they were macrobiological bombs.

The Search for Microbial Life on Mars

The primary objective of the *Viking* mission was to determine if there was evidence of microbial life on Mars. It involved careful planning, sterilization of the craft and the development of precision instruments that could operate robotically and survive the rigours of both the journey and the Martian atmosphere. In addition to minimizing volume, weight and power requirements, the instruments had to operate in a fully automated programme mode, as well as by command from Earth. These technical challenges were daunting, as no one had ever landed a craft on Mars before, so the spacecraft itself also required significant innovative technology.[22]

Viking designers equipped each lander with a 3-metre-long (10 ft) sampling arm and a Biology Package (comprised of three experiments), which had to be shrunk to a mass of less than 15 kilograms (33 lb) and a size roughly equivalent to that of a hatbox; these were the most technically difficult experiments ever attempted by NASA.[23] Each was capable of detecting a wide spectrum of microscopic life within about 12 square metres of the lander. The three biological experiments and their essential findings are summarized below; all yielded significant data in repeated tests on Martian surface samples.

1. *Pyrolytic release experiment.* Headed by Horowitz, this experiment incubated a tiny (0.1 g) soil sample in a test chamber that simulated a Martian atmosphere of carbon dioxide and carbon monoxide; these gases were also labelled with carbon-14 for detection purposes. After five days, the atmosphere was flushed and the sample heated to 625°C (1,157°F) to break down, or *pyrolyze*, any organic material. The resulting gases were passed through a carbon-14 detector to see if any organisms had ingested the labelled atmosphere.
2. *Labelled release experiment.* Developed by NASA's Gilbert Levin, a former sanitation engineer in California who developed a technique to detect bacterial contaminants in water, this experiment moistened a 0.1 g soil sample with 1 cc (cubic centimetres) of a nutrient consisting of distilled water and organic compounds, which had been labelled with radioactive carbon-14. After moistening, the sample was allowed to incubate for at least ten days. It was hoped any microorganisms would consume the nutrient and give off gases containing the carbon-14, which would then be detected.
3. *Gas exchange experiment.* This experiment, popularly known as the 'chicken soup' experiment, was developed by Vance Oyama (1922–1998), chief of NASA's Exobiology Branch at the Ames Research Center. A 0.1 g soil sample was partially submerged in a complex mixture of compounds that the investigators called

> 'chicken soup'. The soil was then incubated for at least twelve days in a simulated Martian atmosphere of carbon dioxide, with helium and krypton added. Gases that might be emitted from organisms consuming the nutrient would then be detected by a gas chromatograph built following a design by the English inventor and scientist James Lovelock (b. 1919), which could detect carbon dioxide, oxygen, methane, hydrogen and nitrogen.

In addition to the three biological experiments, the landers also carried a gas chromatograph mass spectrometer (GCMS), which would look for organic carbon and organic nitrogen compounds. Developed by Klaus Biemann (1926–2016) of the Massachusetts Institute of Technology, the instrument used a thin capillary fibre known as a column to separate different types of molecules, based on their chemical properties.

What were the chances that *Viking* would find evidence of life on Mars? Horowitz placed the chances of his pyrolytic release experiment succeeding as not quite zero. Harold Klein (1921–2001), head of *Viking*'s biology team, gave it a one in fifty chance. And Sagan guessed 50 per cent. A more pessimistic view was held by Lynn Margulis (1938–2011), a microbiologist and member of the National Academy of Science (NAS) Space Science Board's (SSB) Exobiology Subcommittee. In the early 1970s she had been working with Lovelock on a NASA-sponsored search for life on Mars, while also co-developing with him from 1971 the Gaia hypothesis that the Earth can be likened to a living, self-regulating organism. When NASA had tasked Lovelock to invent instruments for its space probes, including some that would help the *Viking* lander in its search for life on Mars, his research led to a profound observation that reflected on his and Margulis's Gaia hypothesis: if life exists on Mars, he argued, sending a spacecraft to the planet was unnecessary, as we can look for it instead by studying the planet's atmospheric

Carl Sagan poses with a model of the *Viking* lander in Death Valley, California.

composition with a spectroscope. As all life tends to expel waste gases into the atmosphere, he said, one could theorize the existence of life on a planet by detecting an atmosphere that was not in chemical equilibrium.[24] His spectroscopic studies of the major gases in the Martian atmosphere, however, revealed no chemical anomalies. The major gases of Mars were in chemical equilibrium. Finding no particular modification attributable to microbes or any other form of life, he and Margulis predicted that the *Viking* probe would find a dead Mars.

The *Viking* results did not settle any bets, as they were ambiguous at best, providing both surprising and contradictory results; while there was no clear evidence for the presence of living micro-organisms in soil near the landing sites, the presence of microbial life could not be totally discounted.[25]

Organic matter was detected in seven of nine runs using the pyrolytic release experiment. Margulis found this a 'very interesting result', adding that 'no one knows how to interpret it'. The data showed an enhanced radioactive signal in both a standard and a control run. 'In some people's minds you've got some sort of positive signal here,' Margulis said. 'In my mind you've got a positive signal that has nothing to do with life at all.'[26] Later attempts to duplicate the effect were unsuccessful.

The labelled release experiment showed an instant – and startling – rise in the level of carbon-14 radioactivity immediately after the nutrient was first introduced into the chamber. A second addition caused an initial drop, then a slow rise, while a control sample gave no release of radioactive gas. These data seemed to attest to a positive reaction, so much so that the experiment team immediately rushed out and ordered a bottle of champagne. Margulis, however, remained unimpressed: 'You can just add chemicals until you release radioactivity,' she said, adding that the results have alternative interpretations: 'neither of them are giving you the kind of complex response you get if you tried to monitor even a desert soil on Earth.'[27] Thus most scientists concluded that the gas was produced by non-living chemistry (brought about by oxidizing agents in the soil).

The gas exchange experiment also produced unexpected results. When the samples were humidified, there was a sudden burst of oxygen (about fifteen times what would have been expected). This was something that had never occurred in earlier tests with terrestrial samples. However, there was a weak response when the nutrient material was added. While the results were dramatic, Margulis notes that 'most people feel that it's a chemical [not biological] response to the adding of the water', adding that, while the release of oxygen is very interesting, and you are going to learn something about the Martian surface, it probably has nothing to do with any kind of living response. 'You see, it stayed constant. It gave off the oxygen and then it stayed constant over quite a long period of time. Which is what happens when you add chemicals and mix.'[28]

The analysed surface samples contained no organic molecules at the parts-per-billion level – less, in fact, than soil samples returned from the Moon by *Apollo* astronauts. Furthermore, the landers' gas chromatograph and mass spectrometer instruments found no sign of organic compounds at either landing site; they did, however, provide a precise and definitive analysis of the composition of the Martian atmosphere and found previously undetected trace elements.[29]

A Stone with Life?

On 27 December 1984 an American meteorite search team chanced upon an interplanetary survivor on the Allan Hills ice field in Antarctica. Known as ALH84001, referring to the fact that it was the first meteorite discovered in the Allan Hills area in 1984, this potato-shaped stone weighing 2 kilograms (4.5 lb) had started its journey to Earth 16 million years ago when an asteroid impacted Mars. The force of the collision blasted rocks off the Martian surface and jettisoned them into space. One orphaned stone wandered through the solar system until 13,000 years ago, when it fell uneventfully onto the Allan Hills ice field. Dated to have an age of 4.5 billion years, ALH84001 has experienced virtually the whole of Martian history and may be part of the Red Planet's initial crust.

A more astounding discovery followed when Christopher Romanek, then working with a NASA–Stanford University team, first noticed that ALH84001 contains some strange-looking carbonate globules no more than 0.25 millimetres across with unusual worm-like structural forms resembling a possible microfossil 1.1 times the width of a human hair. The finding was used in support of a controversial suggestion by some scientists that the meteorite contains microscopic and chemical evidence of ancient life indigenous to Mars, a notion that has not been widely

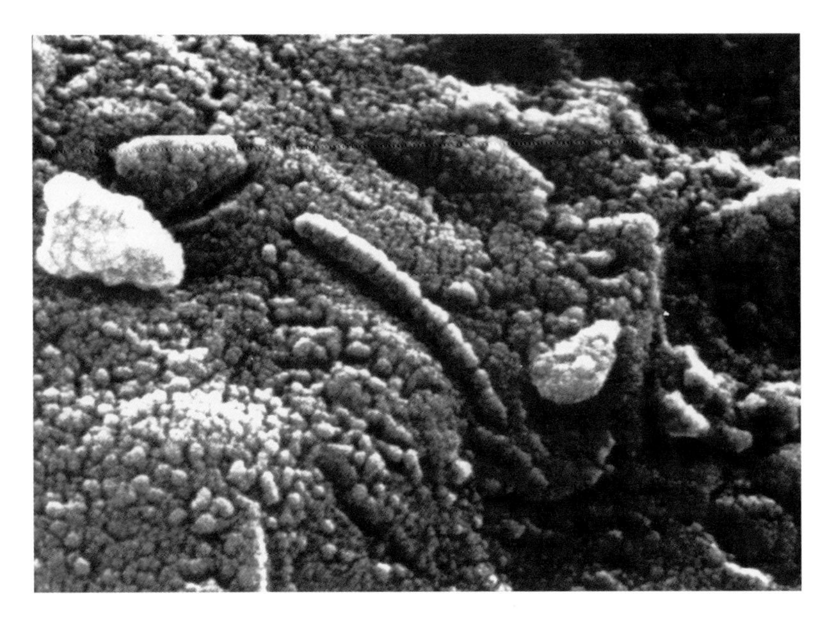

An elongated structure resembling a fossil microorganism (centre of image), revealed in a photomicrograph of a sample of the Martian meteorite ALH84001.

accepted based on the grounds that the forms could be adequately explained by nonbiological processes.

Questions Continue to Grow

Nearly twenty years after the Soviets launched the first satellite into Earth orbit, NASA's twin *Viking* landers became the first fully operational spacecraft on Mars. Armed with a package of biological experiments, they set out to discover if life was present there, but by the end of the mission there was still no answer with any certainty. While the biology results swung the pendulum back to a view of Mars as being a world most likely devoid of life, the other aspects of the mission were a phenomenal success.

By the time the overall mission ended on 21 May 1983, a vast amount of data had been collected: the images from the landers, totalling more than 4,500, and the 52,000 images from the orbiters, which covered 97 per cent of the Martian surface, surpassed expectations in quantity and quality. The landers provided a never-before-seen glimpse of the planet's surface and monitored its atmosphere for changes in clarity and aerosol content over several Martian years. They also sent back images of new and often puzzling terrain, including towering volcanoes and channels through which water may have once flowed, and provided clearer detail on known features, including some colour and stereo observations.

Along with this newfound knowledge came a list of growing questions. If Mars had water in the past, where did it go? Is the planet volcanically active? Are there regions on the planet we haven't explored that may still harbour life? The morsels of uncertainty from the lander's biological experiments alone were enough to keep the question of the possibility of life on Mars open, and the dreams of future missions to Mars alive.[30] But as ambitious and successful as the *Viking* missions were, a tally of the past missions to Mars showed that about half had failed. Was there a better way to explore Mars from space?

An answer came in 1992 with NASA's new administrator Daniel S. Goldin, who 'hated the slow, expensive and not necessarily reliable approach of the past two decades'.[31] Taking hold of the reins, Goldin set out to reform every aspect of NASA. One goal was to reduce the cost of planetary missions from billions of dollars down to $150 million. In a 1992 speech he challenged NASA/JPL to adapt itself to his new 'faster, better, cheaper' philosophy of space flight, which would combine smaller and simpler spacecraft with inexpensive and miniature off-the-shelf components.

A vigorous proponent for increased exploration of Mars, Goldin established a series of robotic missions to visit the planet every two years for the next decade. The missions were developed in one-third the time and at one-tenth the cost of previous Mars expeditions. The programme led to one of the most vibrant and exciting periods of planetary exploration since the 1960s, though it was not without its mishaps.

FIVE

Lifting the Curse at Mars

Following *Viking*, neither the U.S. nor the Soviets had a successful mission to Mars until 1997, though not for the lack of trying. After 1973 the Soviets put Mars exploration on hold as they focused their interests on the Moon, Venus and Comet Halley. Many of these missions were highly successful, so the Soviets tried their luck again at Mars. On 7 and 12 July 1988 they launched their *Phobos 1* and *2* spacecraft, respectively, on an innovative mission to study the Martian moon Phobos that included deploying landers and hoppers onto the surface. *Phobos 1* operated nominally until 2 September 1988, when a communications failure resulted in the spacecraft orientating its solar arrays away from the Sun, thus depleting the batteries. *Phobos 2* was inserted into Mars orbit, where it gathered data on the Sun, the interplanetary medium, Mars and Phobos. Shortly before the craft was to approach within 50 metres (164 ft) of Phobos's surface to release its landers, however, the Soviet curse at Mars struck once more,[1] and a malfunction on the craft's on-board computer caused it to lose contact with Earth.[2] Had these missions succeeded, the Soviets would have been the first to land craft on a moon of another planet.

Less than three years later, the Soviet Union dissolved. Then, in November 1996, Russia made its first post-Soviet attempt to launch a deep-space mission. Called *Mars 96*, this modified *Phobos* craft successfully lifted off, but due to a rocket failure the complex

This image of the Carl Sagan Memorial Station (taken by the *Sojourner* rover) shows the deflated airbags at the base of the spacecraft. The mast for the Pathfinder camera extends upward from the top of the housing. The front rover ramp is perched on top of the airbags and is several metres above the ground. Because of this precarious position, *Sojourner* used the rear ramp to reach the surface. The large rock visible behind the airbags was named 'Yogi'.

spacecraft failed to escape Earth orbit on its way to the Red Planet and plunged into the Pacific Ocean.[3]

NASA also followed the *Viking* success with its own major malfunction. In 1993 *Mars Observer* failed just three days from the craft's planned orbit insertion around the Red Planet; the craft most likely sailed past Mars before entering orbit around the Sun. Originally budgeted at $212 million, *Mars Observer* wound up costing NASA four times as much. As noted in a Jet Propulsion Laboratory (JPL) account of the mission's history, the spacecraft had 'come to symbolize out-of-control costs'. *Mars Observer* became Goldin's favourite target of ridicule and he used it to promote his 'faster, cheaper, better' initiative.

Goldin's philosophy also aligned itself with the NASA budget cuts, which shrank 18 per cent between 1992 and 1999 during the Clinton Administration, which put NASA's JPL on the chopping block. In a speech at JPL on 28 May 1992 Goldin laid all this out for JPL's staff:

> We need to stretch ourselves. Be bold – take risks. [A] project that's 20 for 20 isn't successful. It's proof that we're playing it too safe. If the gain is great, risk is warranted. Failure is OK, as long as it's on a project that's pushing the frontiers of technology.[4]

In 1992 NASA inaugurated a new 'Discovery' programme aimed at producing a series of inexpensive, competitively selected, science-focused missions. The first spacecraft in the new initiative was NASA's *Mars Pathfinder* mission; it cost $150 million, a quarter of the cost of the failed *Mars Orbiter* mission, and was developed in less than three years. *Mars Pathfinder* launched on 4 December 1996 and landed at Mars's Ares Vallis outflow channel on 4 July 1997, from where it began returning the first of its 2.3 billion bits of information.

Artist's visualization of *Pathfinder*'s dramatic airbag landing.

The craft's landing was both unusual and historic. *Mars Pathfinder* was the first spacecraft to use airbags instead of retrorockets to cushion its landing. After impacting the surface at a velocity of about 18 m/s (40 mph), the craft, set in its inflatable cocoon made from a high-strength fibre called Vectran, designed to protect the lander regardless of its orientation upon impact with the surface of the planet, bounced like a ball about 15 metres (49 ft) into the air, and bounced another fifteen times before rolling to rest about 1 kilometre (3,280 ft) from the initial impact site. Once the bags deflated, and the craft revealed itself to Mars, NASA renamed the lander the Carl Sagan Memorial Station (in memory of the *Viking* biology team member, public communicator and long-time proponent for the possibility of life on Mars, who had died on 20 December 1996). The landing was an engineering milestone, as it demonstrated a new way of delivering a spacecraft to the surface of Mars by way of direct entry into the Martian atmosphere. The mission also proved that the concept of 'faster, better and cheaper' missions can work.

Stored within the Station was *Sojourner*, the first rover to reach Mars. Named after Sojourner Truth (1797–1883), a former slave and civil rights crusader who had campaigned for abolitionism and equal rights for women – so it was no coincidence that the craft touched down on Mars on Independence Day – this diminutive six-wheeled robot was the size of a microwave oven and weighed only 11.5 kilograms (25 lb). While the mini-marvel never ventured more than 12 metres (40 ft) away from its lander, and racked up a total of only 100 metres (328 ft) on its odometer, it succeeded in sending back to Earth more than 550 images and fifteen chemical analyses of rocks and soil, as well as extensive data on winds and other weather factors for three months. It also demonstrated for the first time the ability of engineers to deliver a semi-autonomous roving vehicle capable of conducting science experiments to the surface of another planet.

Sojourner rover on the surface of Mars after it exited its carrier (the Carl Sagan Memorial Station) by the craft's rear ramp (seen at lower left). Part of a deflated airbag appears at lower right.

One thought-provoking finding is that the soil chemistry of Ares Valles resembles that analysed at both *Viking* landing sites, but the rocks may have been deposited by floods. Supporting that notion is what appear to be rounded pebbles and cobbles on the ground (and in some rocks), suggesting they belong to conglomerates formed in running water long ago. Of these findings, *Mars Pathfinder* project scientist Matthew Golombek said 'it suggests a water-rich planet that may have been more Earth-like than previously recognized, with a warmer and wetter past in which liquid water was stable and the atmosphere was thicker.'[5] The chemistry of the rocks themselves resemble basaltic andesites (volcanic in composition) and appear different from Martian meteorites found on Earth. According to Golombek, before the *Pathfinder* mission, knowledge of the kinds of rocks present on Mars was based mostly on the Martian meteorites found on Earth, which are all igneous rocks rich in magnesium and iron, and relatively low in silica.[6]

As with the *Viking* landers, *Mars Pathfinder* recorded frequent dust devils, whose gusts may be a mechanism for mixing dust into the Martian atmosphere and turning the sky pink. It also found water-ice clouds forming and abrupt temperature fluctuations in the early morning, implying that the atmosphere is warmed by the planet's surface, with heat convected upward in small eddies.[7]

As *Sojourner* roamed the surface on Mars, it was being followed by an inquisitive public back on Earth. NASA aggressively posted pictures several times a week on its *Pathfinder/Sojourner* website, making use of the Internet when interest in the technology was gaining momentum among the public. In 1997 the Internet was just a few years old: about 36 per cent of adults were online,[8] and 11 per cent of children between the ages of three and seventeen were using the Internet at home.[9] Although dial-up modem speeds achieved only 56 Kbps, anyone with a computer had the ability to monitor *Sojourner*'s progress on another world almost in real time. Had this landing occurred just five years earlier, access to

the images would have only been through universities and government news reports.[10]

NASA's Internet blitz worked, making *Sojourner* an overnight sensation, with the rover making the cover of the 14 July 1997 issue of *Time* magazine, among other accolades. Shortly after the landing, Jet Propulsion Webmaster David Dubov estimated the *Mars Pathfinder* website would be receiving 25 million hits a day. But by day three, the site was up to 80 million hits and going steady. In comparison, the chess match between Gary Kasparov and IBM's Deep Blue computer the year before peaked at 21 million hits and the Atlanta Olympics website had topped out at 18 million hits on one day.[11] *Mars Pathfinder* changed forever how the public expected to get information on NASA missions, and on any other live event.

NASA and JPL lost communication with *Pathfinder* and *Sojourner* on 27 September 1997. Two weeks earlier, however, NASA's *Mars Global Surveyor* (MGS) spacecraft had arrived at Mars, making history as the first spacecraft at the Red Planet to put itself into its final orbit using the technique of aerobraking – brushing against the top of a planet's atmosphere to slow down the spacecraft and lower it into a predetermined orbital altitude. MGS operated in orbit around Mars for nine years and 52 days, longer than any previous spacecraft to Mars.

Among its many discoveries, MGS added to the growing image of Mars not only as a former water-world but one that may still harbour water today. The craft's images provided scientists with several insights into how natural forces sculpt the planet's surface. These included a fan-shaped area of interweaving, curved ridges (interpreted as an ancient river delta resulting from a persistent flow of water over an extended period in the planet's ancient past) and new sediment deposits in two gullies where fluids appear to have carried them down while the craft was in orbit. While the images were the strongest evidence to date that water still flows occasionally on the surface of Mars, if only in brief episodes, some researchers

NASA's *Mars Pathfinder* captured the first colour image ever taken from the surface of Mars showing an overcast sky. The image, taken about an hour and forty minutes before sunrise, shows pink stratus clouds consisting of water ice condensed on reddish dust particles suspended in the atmosphere.

argued that these 'trickle' features more likely have a dry origin, being caused by mini-avalanches of dirt, which can create a similar signature. Furthermore, MGS's mineral-mapping infrared spectrometer found concentrations of fine-grained haematite, a mineral that often forms under wet conditions.[12]

While investigating mapped images that MGS took of the planet from pole to pole, researchers identified twenty new impact craters that had not been present seven years earlier.[13] It is now estimated that Mars gets struck by meteorites about two hundred times a

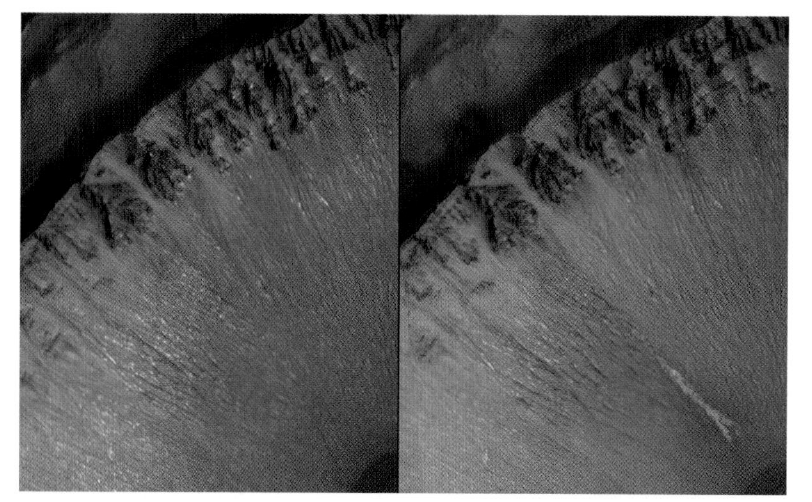

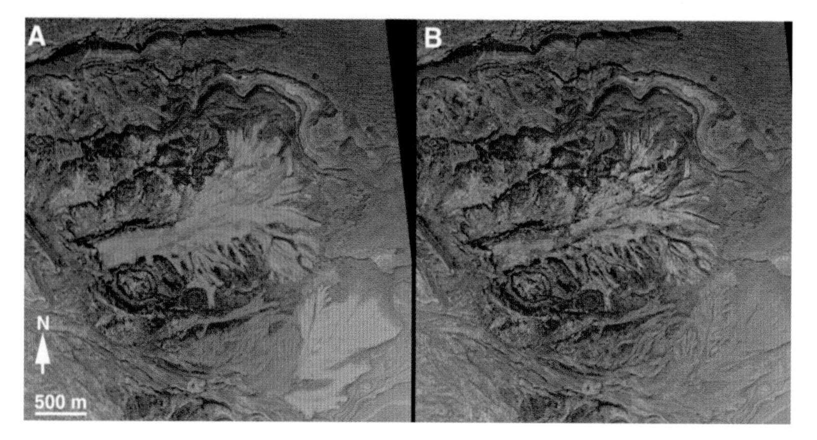

Above: *Mars Global Surveyor* images showing a new gully deposit (light-coloured region in image at right in the Centauri Montes region of Mars, east of the Hellas Basin). The new deposit exhibits characteristics consistent with transport and deposition of a fluid that behaved like liquid water and likely transported some fine-grained sediment along with it.

Below: two fans of sedimentary debris in a surface depression with finger-like protrusions at the ends and sides of the fans. Also preserved are the channels through which water and sediment flowed.

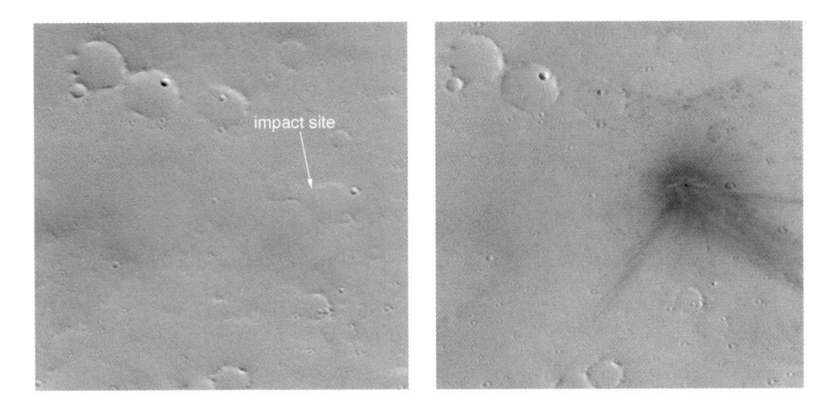

This set of *Mars Global Surveyor* images shows before and after narrow-angle camera views of an impact site. The before image was taken on 24 February 2002. The after image was taken on 13 March 2006.

year. The problem is finding their relatively tiny imprints, roughly 10 metres (33 ft) across.

Other images revealed how carbon dioxide ice deposits near the South Pole were shrinking after three summers due to possible climate change, and hinted that some dark sand dune structures are young and mobile. This had major implications because if the Martian winds are capable of transporting loose sediment and depositing it elsewhere, they must have played a major role in shaping the changing face of Mars over time, explaining what visual and other Earth-based observers have recorded over the centuries. Comparisons of MGS images with *Viking* and *Mariner 9* pictures alone suggest that more than one-third of Mars's surface area has brightened or darkened by at least 10 per cent.[14] NASA lost contact with the MGS orbiter in 2006 after depleted batteries likely left the spacecraft unable to control its orientation,[15] but not before it had revealed the planet to be a rapidly changing world that probably had a different environment in the past.

On 3 July 1998 (JST) the curse at Mars returned when Japan attempted to join the Soviet and U.S. explorations at Mars with the launch of its *Nozomi* (meaning 'Hope') spacecraft. Equipped with fourteen science experiments, *Nozomi* launched successfully, but a series of mishaps and malfunctions made it impossible for it to enter Mars orbit. NASA's *Mars Climate Orbiter* followed with a

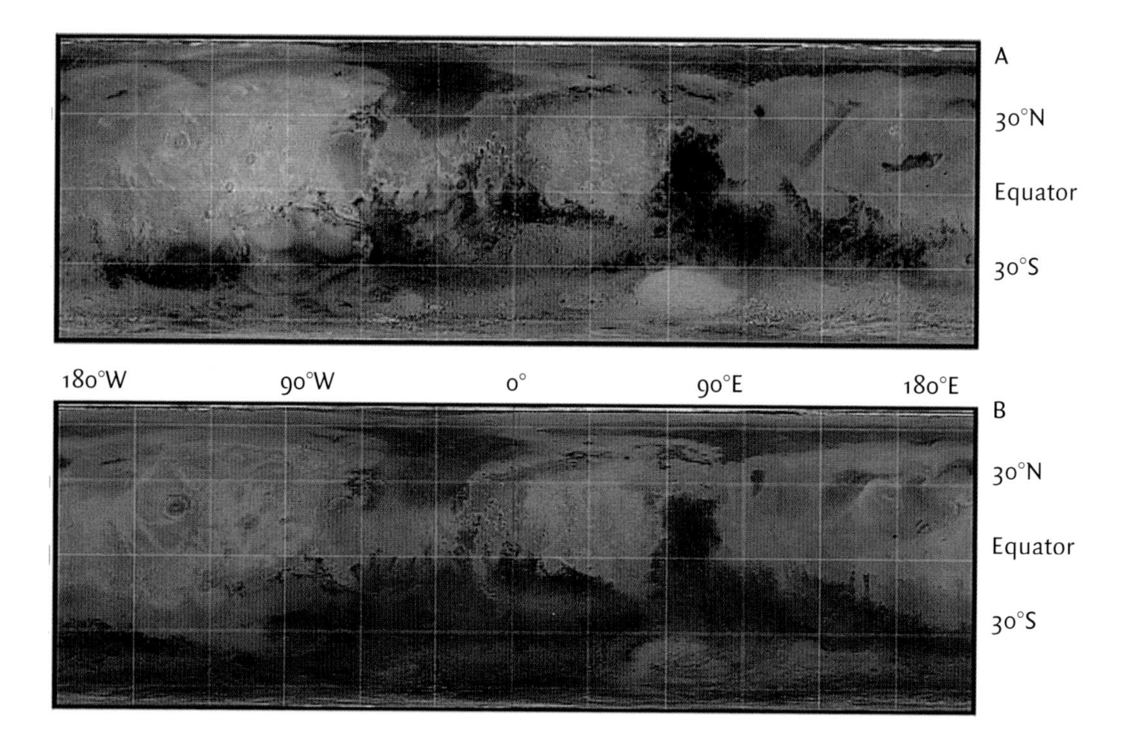

Comparison of *Viking* and MGS global mosaics. (A) *Viking* mosaic created by A. McEwen of the U.S. Geological Survey in Flagstaff, Arizona (1993) using images acquired during the period 1976–80 after the 1977 global dust storms. (B) Cloud-free mosaic of MGS images acquired from 1991–2001 before the 2001 global dust storm, modified from one by P. E. Geissler (2005), also of the U.S. Geological Survey.

successful launch on 11 December 1998, but the mission failed nine months later when the craft burned up in Mars's atmosphere due to a navigation error caused by the failure to translate imperial units to metric. Finally *Mars Polar Lander*, which launched on 3 January 1999, had a premature retrorocket shutdown in December 1999, which resulted not only in the lander crashing but the loss of *Deep Space 2*, two miniature probes that piggybacked onto the *Mars Polar Lander* and were intended to slam onto the Martian surface, but failed to respond to a final communication effort by NASA engineers.[16]

Luck returned to NASA when its *Mars Odyssey* spacecraft arrived at the Red Planet on 24 October 2001 and achieved orbit. After firing its main engine for approximately 22 minutes, the craft allowed itself to be captured into an initial elliptical orbit before it aerobraked into a circular orbit. Exploration at Mars began anew with *Mars Odyssey* being equipped to find evidence for present near-surface water, map mineral deposits from past water activity and monitor radiation levels at Mars as they relate to the potential hazards faced by future astronauts.

As of April 2019 the mission has returned more than 350,000 images and relayed more than 95 per cent of all the data from the robotic *Spirit* and *Opportunity* rovers, which landed on opposite sides of the planet in 2004. *Mars Odyssey* used its Thermal Emission Imaging System (THEMIS) to globally map the amount and distribution of many chemical elements and minerals that make up the Martian surface, a first for any spacecraft at Mars. Among its findings are that olivine-rich materials appear to be widely distributed across the planet. This common green volcanic mineral breaks down rapidly when in contact with water, so its proliferation suggests that some regions of Mars may have been largely dry and cold for billions of years, yielding a vision of Mars as a yin-yang (water/desert) world.[17]

A remarkable diversity of igneous material, both volcanic and plutonic, also exist across its surface. These rocks rival those on Earth and range from low-silica (olivine-rich) crustal basalts typical of fluid Hawaiian-type lavas, which have been observed on crater floors and in layers exposed in deep canyon walls (suggesting they were emplaced at various times throughout the formation of the upper crust), to high-silica (quartz-bearing) dacite and granitic rocks, especially in the Syrtis Major volcanic region.[18]

Collating the spacecraft's data on the distribution of hydrogen on Mars on a map enabled scientists to discover enormous amounts of water ice in the polar regions buried just beneath the surface. These subterranean sheets contain enough water ice to fill Lake Michigan twice over, or enough to cover England from coast to coast – and that may be only the tip of a Martian iceberg.[19]

Together with results from *Mars Global Surveyor*, *Mars Odyssey* has shown that Mars may be emerging from a recent ice age that ended 400,000 years ago.[20] In contrast to Earth's ice ages, a Martian ice age

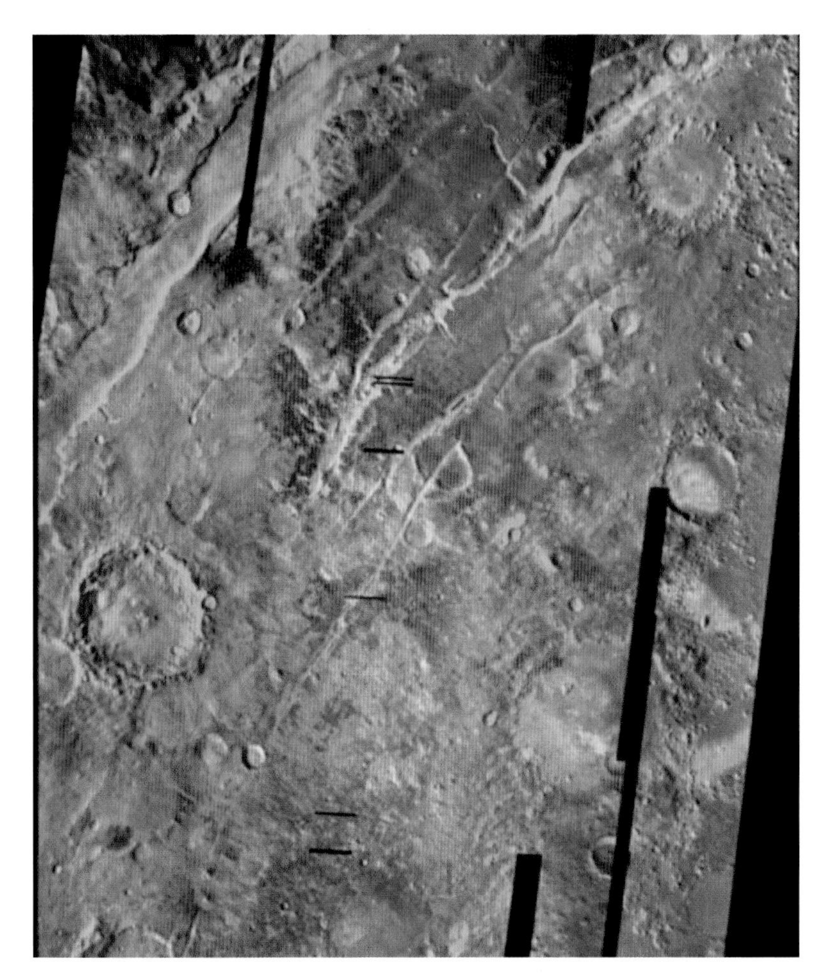

Colours ranging from magenta to purple-blue map the large exposures of olivine-rich rocks in the Nili Fossae region of Syrtis Major. *Mars Odyssey* THEMIS images show the olivine is about four times as extensive as scientists previously thought. This image shows an area about 350 km wide.

waxes when the planet's poles warm up, which transports water vapour towards lower latitudes. They wane when the poles cool and lock water into polar ice caps. Current thinking is that, since the ice on Mars appears to be growing over time, the planet must still be recovering from this icy episode. Understanding the ice caps' Jekyll-and-Hyde behaviour is important for future missions to Mars, so we can plan where the water will be when we send astronauts to the planet. Such knowledge could also potentially help us better understand climate change on Earth.

In December 2003 *Mars Express* – the European Space Agency's (ESA's) first-ever mission to another planet – joined *Mars Odyssey* in Mars orbit with a chief objective to search for water. *Mars Express* derived its name because the spacecraft was built and launched in record time and at a much lower cost than previous, similar missions into outer space. The mission included a *Beagle 2* lander, which the craft released and 'lost' on 25 December 2003 when it arrived at Mars. But images from a later craft (NASA's *Mars Reconnaissance Orbiter*) revealed that *Beagle 2* had indeed made a successful soft landing on Mars, and likely deployed at least three of its four solar panels. Although no one will ever know exactly what happened, it is reasonable to assume that the failure occurred after the landing, as even one disabled solar panel could have seriously hampered *Beagle 2*'s ability to use its antenna, making communication with Earth difficult.

As of late 2019, both *Mars Odyssey* and *Mars Express* continue to deliver a daily stream of extraordinary images and scientific data, with *Mars Express* having captured thus far more than 21,000 images and made several important discoveries relating to Mars as a

False-colour THEMIS map of Mars. Soil enriched in hydrogen is indicated as deep blue. Note the prevalance of the colour in the polar region.

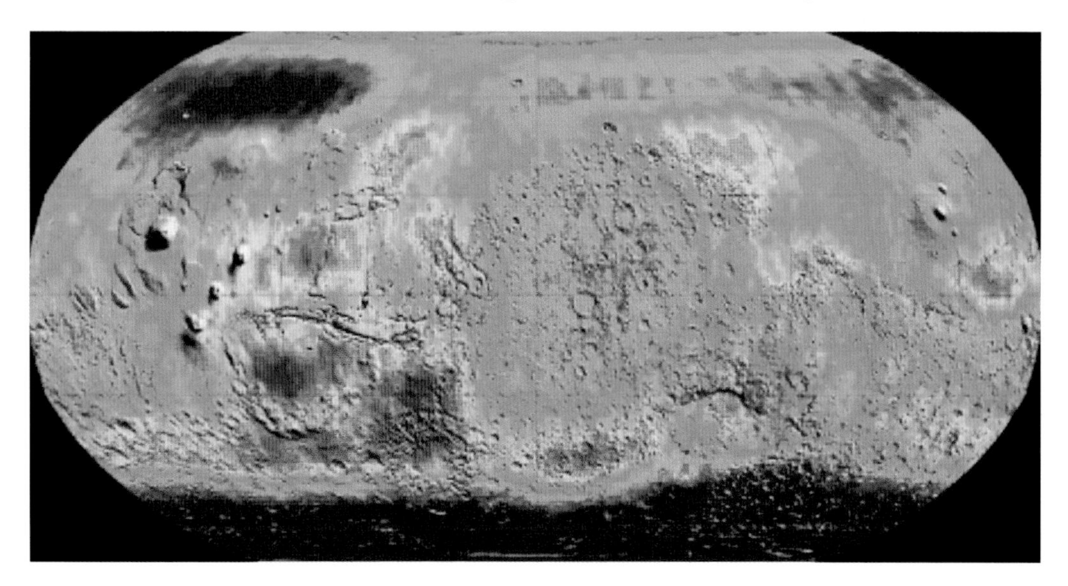

water-world. The craft's radar sounder (MARSIS), the first to orbit another planet, quickly detected subsurface layers of water ice, complementing the findings by *Mars Odyssey*. Another instrument, OMEGA, a combined camera and infrared spectrometer, revealed vast plains of permafrost around the planet's South Pole. One outcome of these studies is that the total amount of water ice contained at the South Pole of Mars makes up the largest water reservoir on the planet today. If this polar ice melted, the entire surface of the planet would be covered by an ocean 11 metres (36 ft) deep.[21]

While investigating the planet's atmosphere, *Mars Express* detected the spectral fingerprint of a tiny amount of methane gas (about ten parts in a thousand million). Unless methane is continuously produced by a source, it would only survive in the Martian atmosphere for a few hundred years, because it quickly oxidizes to form water and carbon dioxide, both present in the Martian atmosphere. So there must be a mechanism that refills

The bright white region of this image shows the icy cap that covers Mars's South Pole.

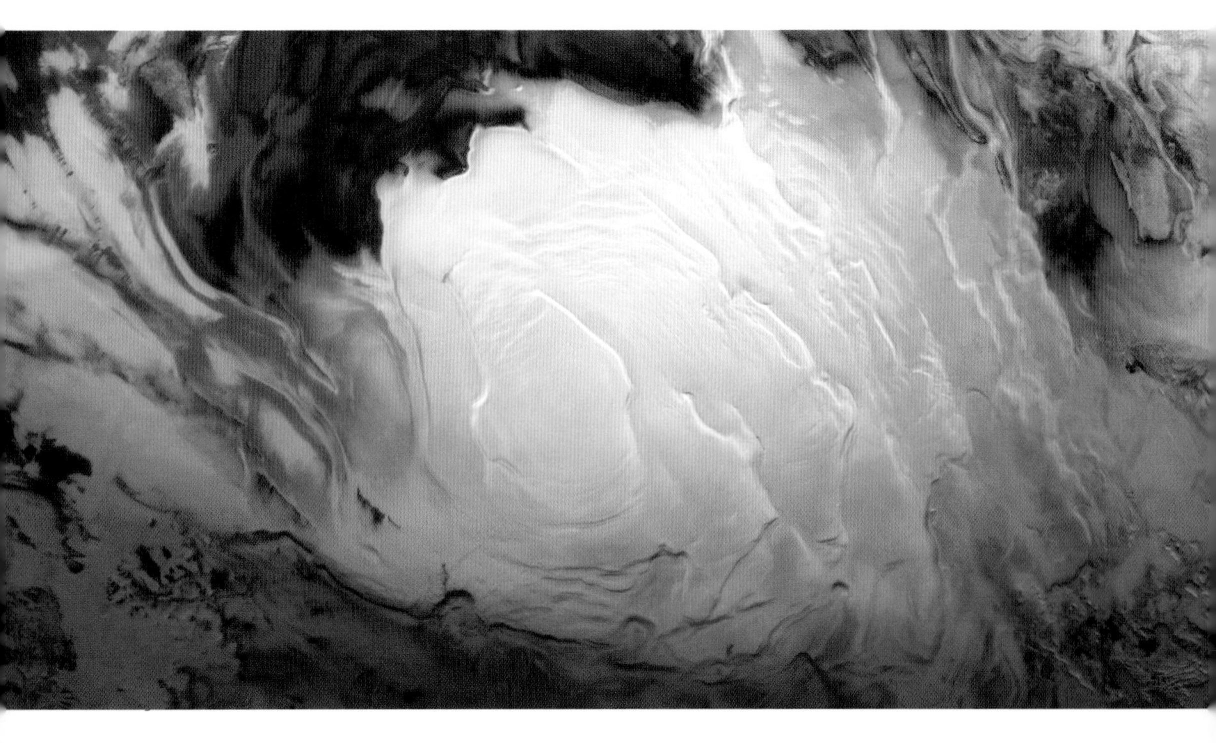

This perspective view in the Colles Nili region on Mars was generated from the high-resolution stereo camera on ESA's *Mars Express*.

the atmosphere with methane: the source could be linked to either volcanic or geothermal activity on Mars, or perhaps biological activity, such as fermentation.[22]

The images taken by *Mars Express* are among the most remarkable to date. The craft has revealed many 3D landscapes in perspective to give us a more realistic view of the planet. One of the most exciting discoveries was evidence of recent and recurring glacial activity in tropical and mid-latitude regions. For instance, the images revealed a stunning terrain in the Colles Nili region at the boundary between the Red Planet's southern and northern highlands. One of the oldest and most prominent features on Mars, it marks the location of a height difference of several kilometres between the two hemispheres.

Images of this feature show a jumble of eroded blocks, most likely erosional remnants of a former plateau, with remnants of ancient glaciers flowing around them.[23] Both the layered deposits and the ridges and troughs are thought to be associated with buried ice that has since been covered over by wind-blown dust and local debris from the eroding plateau, perhaps as an underlying ice sheet retreated.

These observations indicate that glaciers formed until only a few million years ago, when the planet was warmer and possibly also had a thicker atmosphere. The climate then seems to have changed, possibly as a result of a shift in the tilt of the planet's polar axis, so that the glaciers became inactive or retreated due to the lack of a continuing supply of water. Since then they have been protected from sublimation by a thin coating of dust.

Another important discovery was revealed in a survey of the Hephaestus Fossae region. This is a long and intricate channel network (indicative of the Red Planet's watery past) associated with a crater 20 kilometres (12.5 mi.) wide. Most likely the soil – a mixture of rock, dust and subterranean water ice – melted when a comet or asteroid hit the planet, resulting in the fluidized appearance of the impact's debris blanket that surrounds the crater. Based on the lack of similar structures near smaller nearby craters, it appears that only impacts by sizeable objects dig deep enough to release part of the frozen reservoir of water lying beneath the surface.

This false-colour *Mars Express* image shows a 20 km crater and fluidized flow in the Hephaestus Fossae. Green and yellow shades represent shallow ground, while blue and purple show deep depressions, reaching depths of 4,200 m.

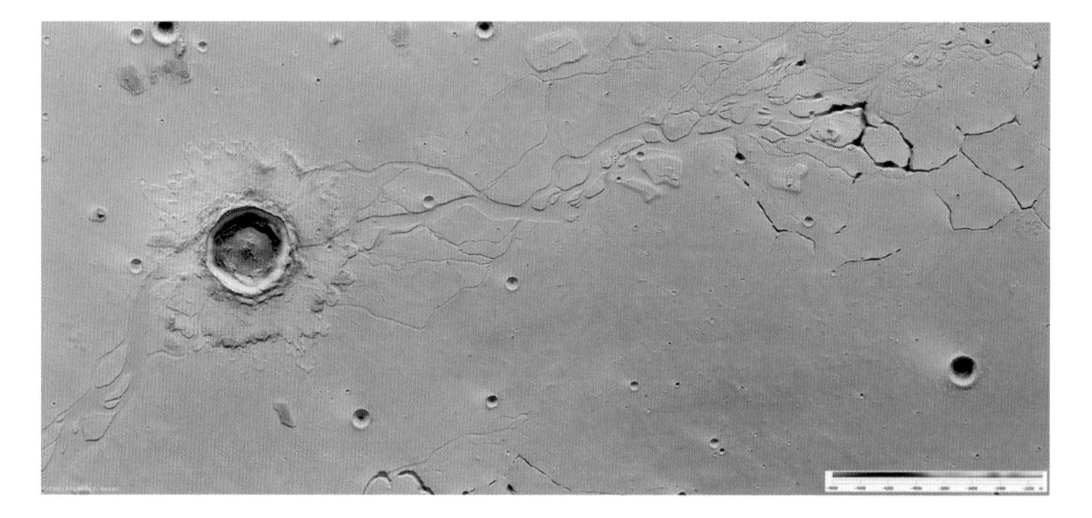

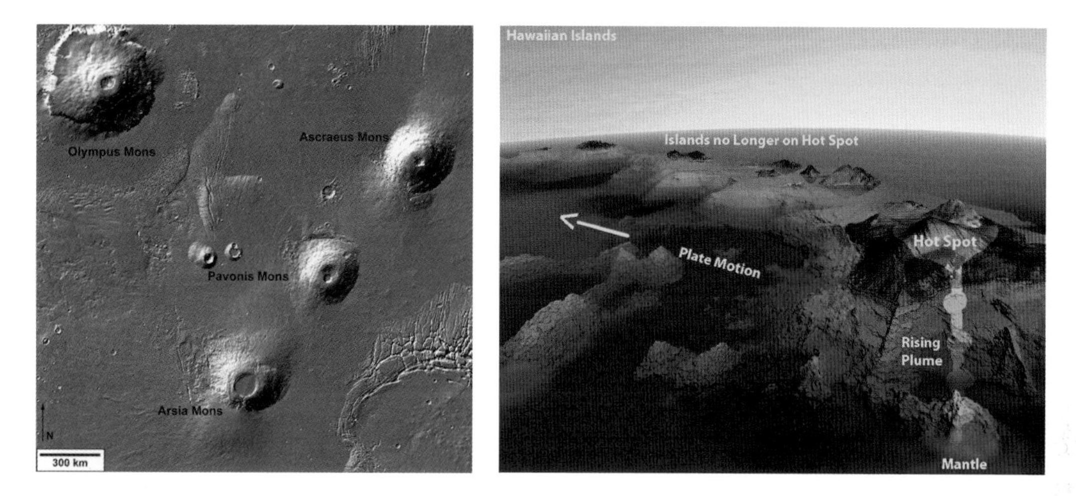

Left: NASA's *Mars Global Surveyor* shaded relief image of the three major volcanoes of Tharsis Montes – Arsia, Pavonis and Ascraeus Mons – and solitary Olympus Mons. New *Mars Express* data suggest these constructs formed one by one, starting with Arsia Mons, possibly by a single mantle plume moving under the surface.

Right: while the Tharsis chain appears similar to the Hawaiian island chain, they formed differently. The Hawaiian islands likely form as the plate on which they lie moves over a stationary plume of rising magma. The Martian volcanoes, on the other hand, may have formed as a migrating mantle plume moved underneath a stationary crust.

In February 2019, ESA announced a novel discovery by *Mars Express*: the first geological evidence of a system of interconnected lakes whose water level aligns with the proposed shorelines a putative Martian ocean thought to have existed some 3.5 billion years ago. Gian Gabriele Ori, director of the Università D'Annunzio's International Research School of Planetary Sciences in Italy, notes that the ocean may have connected to this system of underground lakes that spread across the entire planet, making them contemporaries of a Martian ocean. The finding is hugely significant as it will help future missions to Mars pinpoint the most promising locations for finding signs of past life.

In addition, five years of *Mars Express* gravity mapping data indicate that lava in the planet's Tharsis bulge originated as a lighter andesitic lava, which can form in the presence of water, before becoming overlaid with the heavier basalts we see on the surface today. The Tharsis basalts also resemble the composition of meteorites that have fallen to Earth.

The data, combined with the varying height of the volcanoes, point to Arsia Mons being the oldest, followed in sequence by Pavonis and Ascraeus Mons. As no evidence for plate tectonics has been found in the region, it appears that a drifting plume of molten

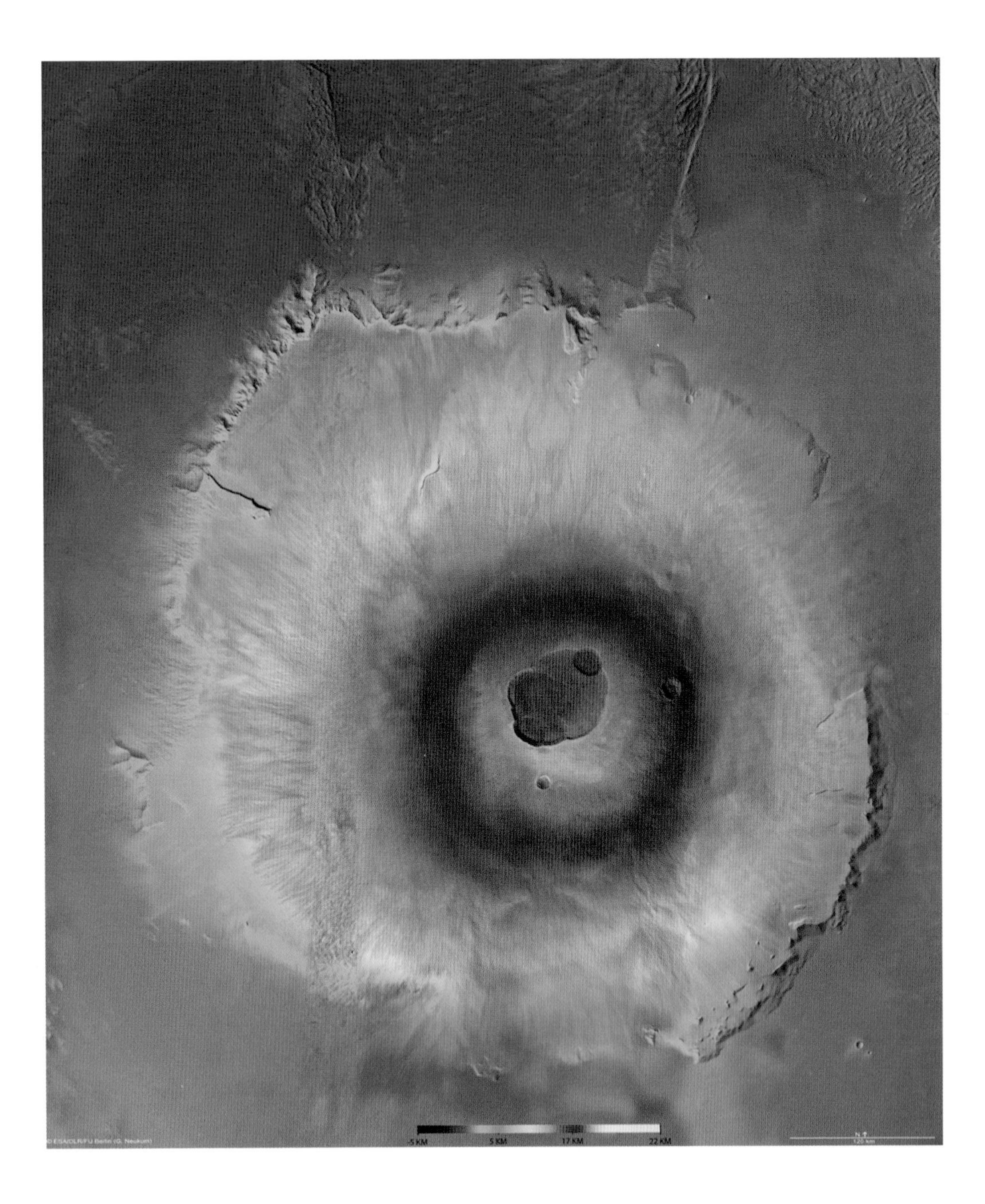
© ESA/DLR/FU Berlin (G. Neukum)
-5 KM
5 KM
17 KM
22 KM
N
120 km

Olympus Mons colour-coded according to height from white (highest) to blue (lowest). The data are based on images captured by the High Resolution Stereo Camera on board ESA's *Mars Express*.

rock beneath the surface is responsible for this progression. The scenario is reminiscent of the hot spot volcanism that sequentially formed the Hawaiian volcanoes; the difference, however, appears to be that the Hawaiian islands formed as the Pacific oceanic plate moved over a stationary molten mantle plume (likened to a blob in a lava lamp but on a gigantic scale), while the Tharsis Montes volcanoes may have been formed by one or more molten mantle plumes, slowly moving laterally beneath the surface.[24]

Olympus Mons, on the other hand, sits on the edge of the Tharsis bulge on a region of crust more rigid than that found on the Tharsis bulge itself, as researchers discovered from the *Mars Express* data. It appears that the Tharsis volcanoes partially sank into a less rigid crustal foundation. This difference between Tharsis and Olympus Mons could suggest great differences in temperature across the plume as it formed the volcanoes. The greater crustal thickness of the bulge may have insulated the inner molten rock leading to high temperatures (and thus a less rigid support structure), while magma below Olympus Mons probably ascended through older and colder crust, which acted as a more rigid foundation.

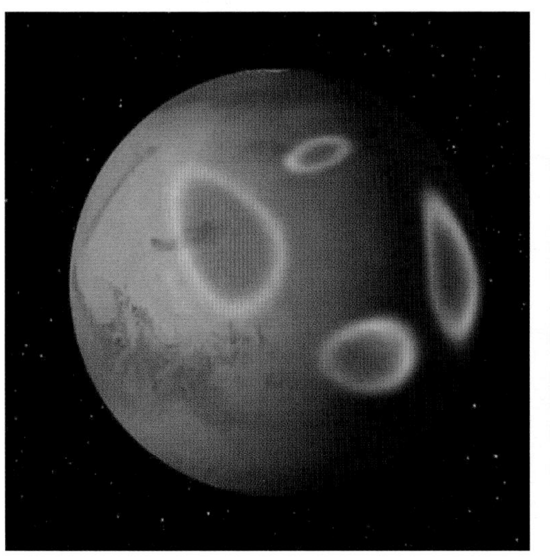

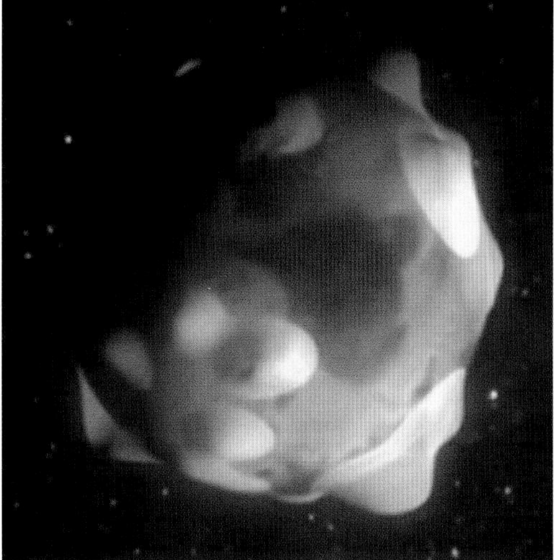

As discovered by *Mars Express*, Mars has magnetized rocks in its crust (left) that create magnetic fields that extend into space above the rocks. When these open fully like an umbrella (right), they emit detectable ultraviolet radiation.

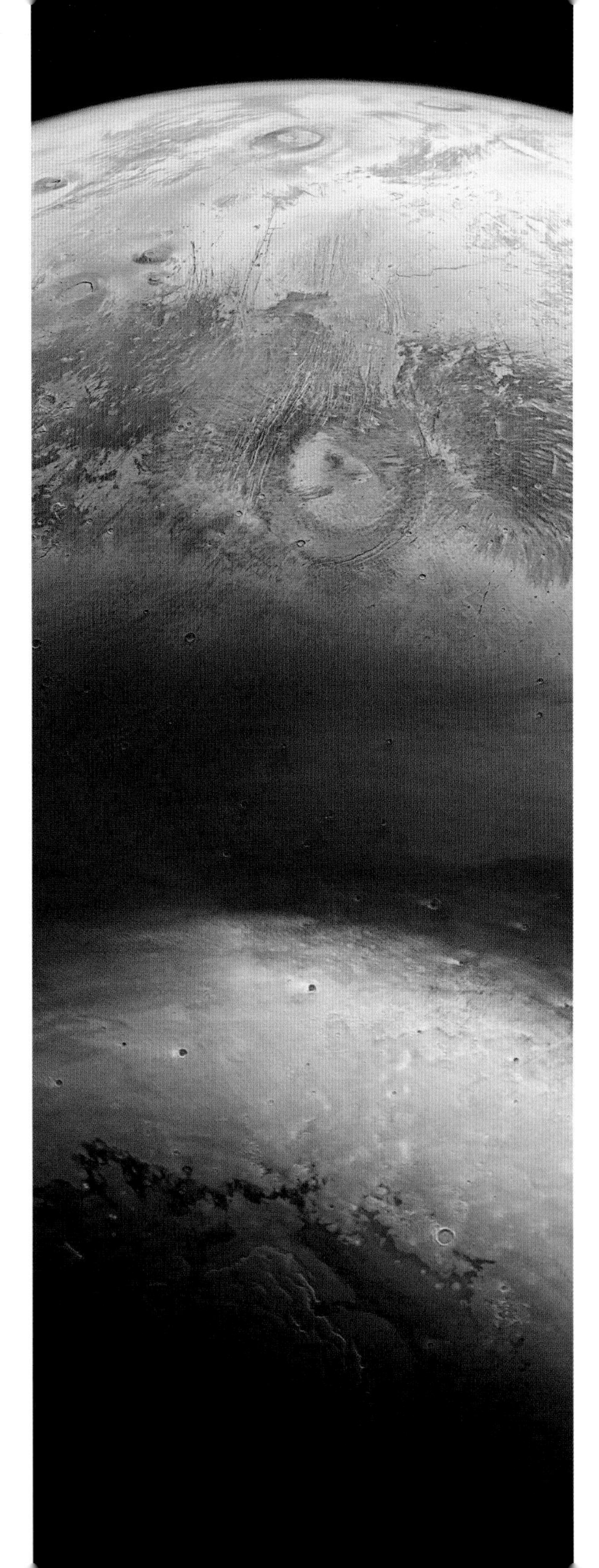

This stunning image swath was taken on 19 June 2017 by ESA's *Mars Express* during camera calibration as the spacecraft flew over the North Pole (bottom) towards the equator (top).

Mars Express perspective image showing that a flat-topped mesa is located in the centre of Hebes Chasma, an enclosed trough, almost 8,000 m deep, in Valles Marineris, the Grand Canyon of Mars, where water is believed to have flowed.

In its observations to look for anomalies in the crystal magnetic field, *Mars Express* made the first detection of a Martian aurora. In August 2004, the spacecraft imaged rare localized ultraviolet auroras linked to magnetized rocks in the planet's crust. Unlike Earth, Mars does not have a powerful global magnetic field, so its atmosphere is devoid of any dramatic colourful displays of northern or southern lights. However, *Mars Express* data showed that the Red Planet is instead peppered with pockets of magnetism linked to magnetized rocks in its crust, each potentially capable of connecting with the wind of particles from the Sun to spark modest auroral displays in ultraviolet light.[25]

A decade after that discovery, ESA researchers realized that Mars's ultraviolet auroras appear to be not only a rare and transient phenomenon but very different from what is observed on other planets. Auroral displays on Mars are very short-lived, do not appear to repeat in the same locations and only occur near the boundary between open and closed magnetic field lines. The emission is detectable only when the local magnetic field opens like an umbrella,

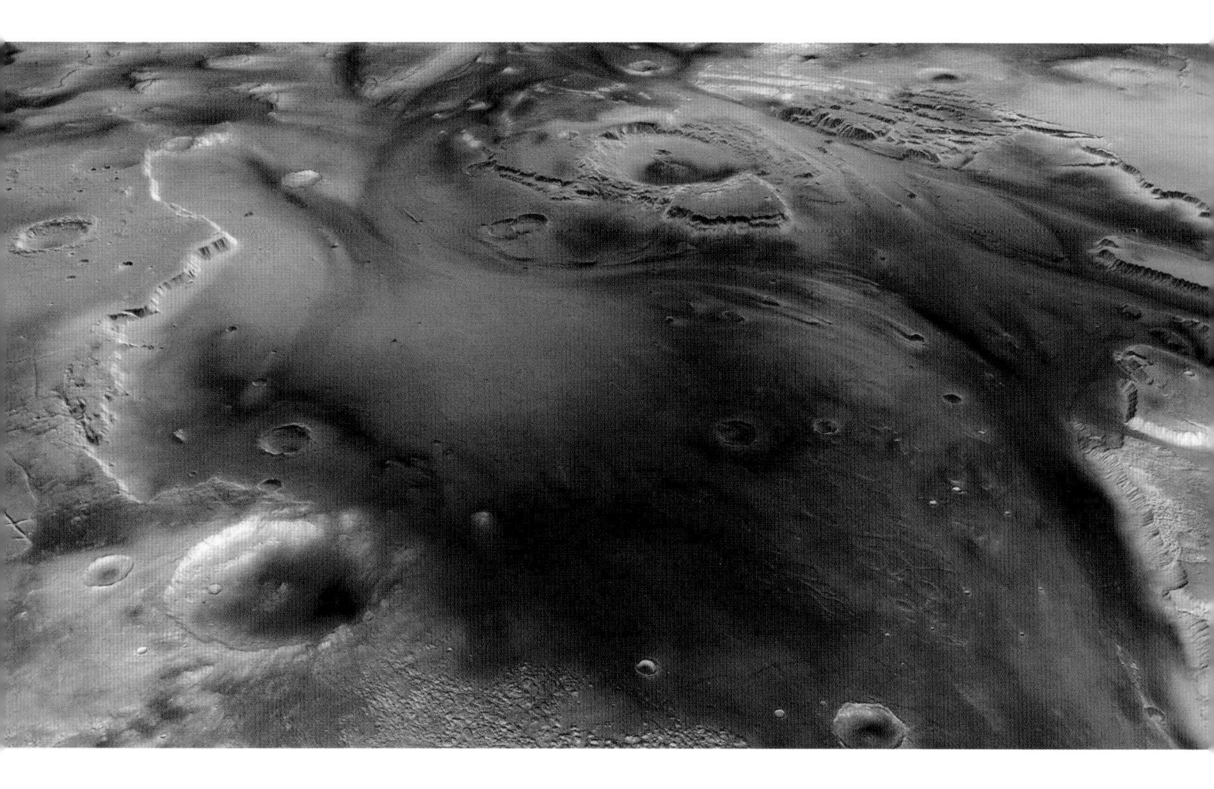

Mars Express perspective image looking in the direction of once-flowing water from the northern flanks of Kasai Valles towards the 100-km-wide Sharonov crater in the top centre, and with Sacra Mensa to the lower right.

allowing access to the energized electrons at the tips of these magnetic domes. These intriguing aurorae were to become key targets for later spacecraft missions to Mars with hopes of unlocking their secrets.

Studies of Mars's magnetic environment are also key to understanding the early fate of the planet's atmosphere, which is leaking into space. Mars had a magnetic field shortly after it was created some 4.5 billion years ago. At that time, young Mars had a churning molten iron core, a magnetic field and a dense atmosphere, just as Earth does today. But sometime around 4.2 billion years ago Mars's core and the planet lost much of its atmosphere (which is now only about 1 per cent as thick as Earth's) and its magnetic field. Understanding why Mars has lost much of its atmosphere, especially early in its history, which transformed the planet from a warm and wet world to a cold and arid place, is essential to understanding what makes a planet habitable.

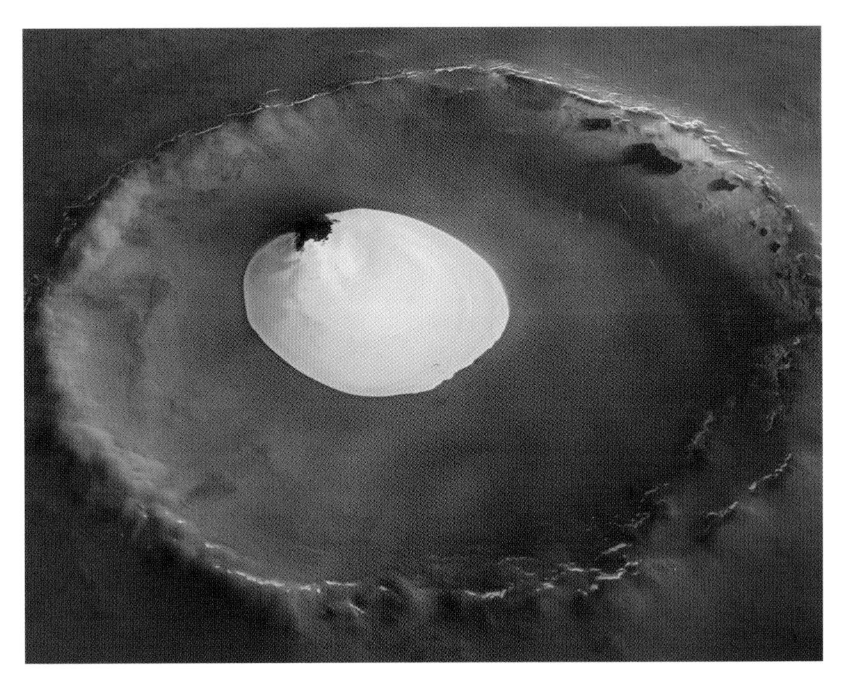

Mars Express image showing a 35-km-wide crater (with a maximum depth of about 2 km beneath the crater rim). The circular patch of bright material located at the centre of the crater is residual water ice.

For years researchers believed that the planet's low gravity and lack of a magnetic field made its outermost atmosphere an easy target to be swept away by the solar wind.

As announced in February 2018, however, after fourteen years of monitoring the Red Planet's atmosphere for charged ions, such as oxygen and carbon dioxide, flowing into space, *Mars Express* uncovered a surprising effect: increased ion production in the Martian atmosphere, triggered by ultraviolet solar radiation, actually shields the planet's atmosphere from the detrimental energy carried by the solar wind. However, as Robin Ramstad of the Swedish Institute of Space Physics, and leader of this *Mars Express* investigation, explained, 'very little energy is actually required for the ions to escape by themselves, due to the low gravity binding the atmosphere to Mars'.[26] Thus the Sun's ultraviolet radiation appears to be playing a more important role than previously thought.

The Magic of Roving Mars with 'Spirit' and 'Opportunity'

Only one month after *Mars Express* arrived at the Red Planet, NASA safely landed twin rovers (*Spirit* and *Opportunity*) on the Martian surface. *Spirit* landed in the crater Gusev, which was once filled with a lake. *Opportunity* landed in Meridiani Planum, an area once thought to have been covered in water. During its mission, *Spirit* imaged dust devils moving across the surface of Mars and Earth in the Martian sky, the first ever taken from the surface of another planet. *Spirit* put nearly 8 kilometres (5 mi.) on its odometer of Martian terrain before communications were lost on 22 March 2010, far outliving its planned ninety-day mission. *Opportunity* also embarked on what was to be a ninety-day mission, but the

Spirit took this image of its tracks in Martian soil. The wheel at the bottom right of the image is *Spirit*'s right-front wheel. Because that wheel no longer turned, *Spirit* drove backwards.

The small spherules (blueberries) in this close-up *Opportunity* image are near Fram Crater. The area shown is 3 cm across.

six-wheeled rover kept on rolling until it stopped communicating with Earth when a severe Mars-wide dust storm blanketed the region in June 2018; after more than a thousand unanswered commands sent to the rover to restore contact, NASA officially ended the mission on 13 February 2019.

By the time *Spirit* and *Opportunity* landed on Mars in 2004 NASA had moved its web infrastructure into a commercial data centre and added a commercial caching network. The change allowed NASA to handle even more traffic, achieving 109 million hits in 24 hours, including the 50,000 people who watched NASA TV's coverage of the landings via webcast.

Together, the twin rovers have sent to Earth hundreds of thousands of spectacular, high-resolution, full-colour images of the Martian terrain as well as detailed microscopic images of rocks and soil surfaces. Four different spectrometers have amassed unparalleled information about the chemical and mineralogical make-up of Martian rocks and soil. Special rock abrasion tools, never before sent to another planet, have enabled scientists to peer beneath the dusty and weathered surfaces of rocks to examine their interiors.

Both rovers found ample evidence for past wet conditions that possibly could have supported microbial life: *Spirit* discovered rocks with sulphate salts deposited by water, and *Opportunity* found ripple marks and layered bedrock in the remains of a shallow, ancient sea. *Opportunity* also discovered BB-sized (that is, shot pellets 0.18 in. in diameter) heamatite-rich 'pellets' (nicknamed blueberries) of water-formed rock.

Since 1996 missions to Mars have undergone a Renaissance, employing revolutionary new ways to orbit and land on the planet. Data from the successful orbiter and lander missions have provided scientists with an equally revolutionary new view of Mars as a once Earth-like world with a complex geologic history. A new Mars was rising, one whose image harked back to the early days of Mars exploration when the planet was envisioned as a world that has conditions that still could nurture life. The latest Mars explorers also paved the way for further scouting missions, including programmes that would take advantage of new and creative technologies to offer greater longevity and mobility on the surface. Many exploring strategies also took on a new focus: to search for and visit regions that suggest the best case for safeguarding present or preserving evidence of past life, while scouting out the safest locations for human exploration of Mars.

SIX

SCOUTING OUT HUMANITY'S NEXT HOME

NASA's success continued with the launch of its *Mars Reconnaissance Orbiter* (MRO) on 12 August 2005. After a seven-month cruise to Mars and six months of aerobraking to reach its science orbit, the 2,180-kilogram (4,805 lb) craft, the largest orbiter sent to Mars in thirty years, almost immediately began collecting data on its four principal objectives:[1]

1. *Determine whether life ever arose on Mars.* As all life (from microbes to more complex organisms) require water to exist, MRO would zero in on water-related mineral deposits and surface features, such as outflow channels from ancient floods.
2. *Characterize the climate of Mars.* By searching for and identifying present-day ice or liquid water beneath the surface, MRO would explore the subsurface structure of the polar caps and nearby terrain for any evidence of the role water may have played in Mars's past climate.
3. *Characterize the geology of Mars.* Beyond the search for signs of active geology, such as volcanism, MRO would particularly focus on geologic settings that might reveal the presence of liquid water on the surface at some point in the planet's history, including the folded layers of Mars's surface; like tree rings, these features keep a geological record of Mars's history.

4. *Help prepare for human exploration.* MRO imaging and other data would help scientists on Earth scout out the most promising and safest locations for scientific study, which would help pave the way for future robotic missions and prepare for sending humans to Mars.

To achieve these goals, the orbiter carries six instruments with wide-ranging capabilities, surveying Mars from its underground layers to the top of its atmosphere. Chief among them are its Context Camera, which can resolve surface features smaller than a tennis court; the Compact Reconnaissance Imaging Spectrometer, capable of providing information on the composition of the Martian surface at scales of 100 to 200 metres (330 to 655 ft) per pixel; and its High Resolution Imaging Science Experiment (HiRISE), the most powerful telescopic camera ever sent to another world, which can reveal rocks the size of a small desk from 300 kilometres (186 mi.) above the surface. In addition, an advanced mineral-mapper can identify water-related deposits in areas as small as a large living room carpet, and there are radar probes for buried ice and water, a weather camera and an infrared sounder (for recording temperatures and atmospheric water vapour concentrations) for monitoring Mars daily.

One of the most celebrated findings was a discovery announced in January 2018 of eight sites where thick deposits of water ice lie exposed under roughly a third of the Martian surface, in the faces of eroding slopes. Shane Byrne of the University of Arizona Lunar and Planetary Laboratory in Tucson said, ‘It’s like having one of those ant farms where you can see through the glass on the side to learn about what’s usually hidden beneath the ground.’

MRO’s data of the ice reveals it was likely deposited long ago as snow before it was capped by a layer 1 to 2 metres (3–6.5 ft) thick of ice-cemented rock and dust. These sites not only hold clues about Mars’s climate history, but may make frozen water more accessible

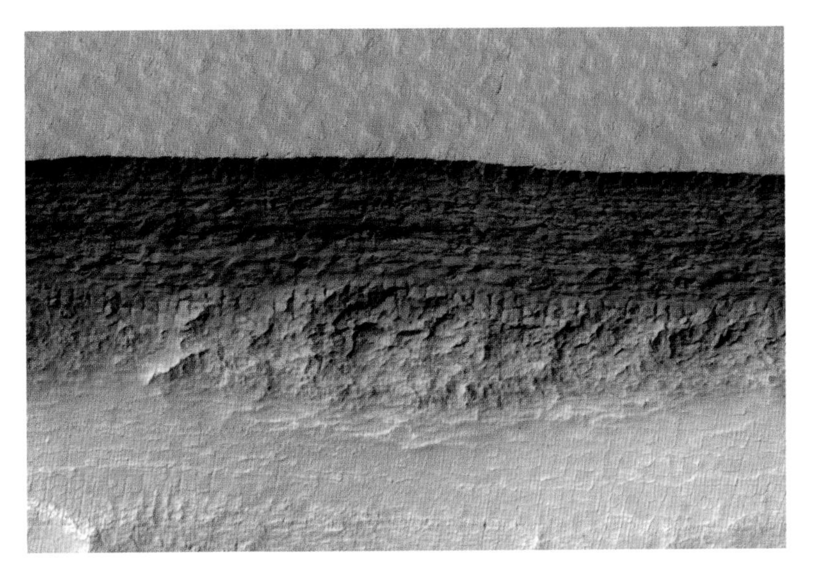

A cross-section of underground ice is exposed at the steep slope that appears bright blue in this enhanced-colour view from the HiRISE camera on NASA's *Mars Reconnaissance Orbiter*. The scene is about 550 m wide. The scarp drops about 140 m from the level ground in the upper third of the image.

than previously thought to future robotic or human exploration missions. As these sites lie at altitudes with less hostile conditions than at Mars's polar ice caps, Byrne said that 'astronauts could essentially just go there with a bucket and a shovel and get all the water they need'.[2]

In another surprising finding, MRO data provided supporting evidence for past explosive volcanism on Mars, revealing patches of iron oxides and sulphates in a region with flat-topped mountains known as Sisyphi Montes. These mountains resemble volcanoes on Earth that have erupted underneath ice. Indeed, the iron oxide and sulphate minerals detected by MRO can only collectively form on the surface from one type of volcanic eruption: a subglacial eruption. Explosive volcanism does not necessarily require silica-rich magma (such as those like Mount St Helens, Washington, on Earth, which Mars does have), but rather, as the eruption of Eyjafjallajökull in Iceland in 2010 exhibited, a very runny, high-temperature basaltic Hawaiian-type lava emerging from a fissure that mixes with overlying glacial ice to generate enormous, sustained ash columns.

MRO image of mounds, hundreds of metres in size, scattered throughout Chryse Planitia. The largest of the mounds show a central pit, similar to the collapsed craters found at the summit of some volcanoes on Earth. While the origins of these pitted mounds or cratered cones are uncertain, they could be the result of the interaction of lava and water, or perhaps formed from the eruption of hot mud originating from beneath the surface and may be one of the long-sought-after sources for transient methane on Mars.

Such findings have implications in the search for life on Mars. University of London scientist Claire Cousins, who has studied volcanic environments in Iceland and Antarctica as Martian analogues, and is investigating subglacial volcanoes on Earth as habitats for life, said that regions of volcano–ice interactions on Earth 'provide a wide range of hydrothermal environments that can be exploited by microbial life'.[3]

Adding to the new views of a wetter Mars, MRO transmitted evidence of a hydrated form of silica on the Martian surface similar to opal on Earth.[4] The presence of opal in these relatively young rocks tells scientists that water, possibly as rivers and small ponds, interacted with the surface as recently as 2 billion years ago, 1 billion years later than scientists had previously believed. The discovery supports other findings that water played an important role in shaping the planet's surface in the past and possibly hosted life. Other science observations have revealed dynamic activity on today's Mars, including the presence of fresh craters, avalanches, seasonal freezing and thawing of carbon dioxide sheets, summer-time seeps of brine and dust storms, including the great global storm of 2018.

MRO remains operational as of November 2019. The craft has completed more than 60,000 orbits around Mars. After fourteen years of collecting data, the spacecraft has taken more than 378,000 images and returned more than 361 terabits of data. NASA announced in February 2018 that it plans to operate the spacecraft beyond the mid-2020s, an extraordinary feat given that the mission was initially designed for two years at Mars. MRO continues to help scientists appreciate a world that has changed greatly over time – a story told, in part, by glacial ice deposits.

Over hundreds of thousands of years, Mars, much like Earth, has undergone variations in its tilt and the shape of its orbit. Radar data from MRO gathered around the northern polar ice cap support previous findings by *Mars Global Surveyor* and *Mars Odyssey* that the most recent ice age on Mars ended about 400,000 years ago.[5] Models suggest that since then the polar deposits should have thickened by about 300 metres (985 ft), which matches the 320 metres (1,050 ft) shown in MRO images, equivalent to a 60-centimetre-thick (24 in.) global layer of ice. Using these measurements, scientists can improve our understanding of how much water is moving between the poles

Climatic cycles of ice and dust build the Martian polar caps, season by season, year by year, and periodically whittle down their size when the climate changes. MRO has revealed a record of the most recent Martian ice age recorded in the planet's north polar ice cap, as simulated here in a 3D perspective based on *Mars Odyssey* image data.

and other latitudes, helping to locate areas of interest to future astronauts on Mars.

These and similar MRO findings over fourteen years of operation have revealed in unprecedented detail a planet that held diverse wet environments billions of years ago and remains dynamic today. Of the seven current Mars missions still in progress at the end of 2019, MRO returns more data every week than the other six combined. Thanks to the mission's longevity, and to the productivity of the orbiter's instruments, MRO still has a major role to play as an advance scout for both rover and human missions.

MRO image of layered sediments in Danielson Crater. These layered sediments are of great interest because they are very regular in thicknesses, suggesting some sort of periodic process such as climate change associated with Mars orbital variations.

Phoenix on Ice

NASA augmented its study of the history of Martian water with the 25 May 2008 landing of *Phoenix*, part of NASA's Scout programme (an initiative for smaller, lower-cost spacecraft). In August 2008 the lander completed its three-month mission (plus two extra months) during which time it studied the high northern arctic plains and atmosphere of Mars. It landed farther north than any previous mission, at a latitude (between 65° and 75°N) equivalent to that of northern Alaska on Earth. Designed to probe the history of water that may have existed in the arctic as recently as 100,000 years ago, the lander used a robotic arm, dug several trenches and collected samples of both soil and water for analysis in its tiny chemistry lab for the chemicals of life. Consequently *Phoenix* became the first Mars lander to identify water in a soil sample. As William Boynton, lead scientist for the craft's Thermal and Evolved-Gas Analyzer, said, 'this is the first time Martian water has been touched and tasted'.[6]

An unexpected surprise finding from *Phoenix* was that it does snow on Mars and that water ice in the area sometimes thaws enough to moisten the soil, which contains calcium carbonate. As with the *Viking* landers, *Phoenix* scooped up and analysed soil samples in a chemistry lab, but it is uncertain whether the soil thawed enough to be available for carbon-based organic compounds to be present, especially as the perchlorate salts detected in the soil samples could have broken down simple organic compounds during the heating of the samples in its lab, making the results inconclusive.

The discovery of perchlorate salts was one of the mission's biggest revelations. This 'multi-talented' chemical strongly attracts water, and can pull humidity from the Martian air, or at higher concentrations combine with water as a brine that stays liquid at Martian surface temperatures. Some microbes on Earth use perchlorate as food. Human explorers might find it useful as rocket fuel or for generating oxygen.

The *Phoenix* lander completed its three-month mission in August, but NASA officially ended the mission after its last signal was received on 2 November 2008. NASA declared the lander officially dead in May 2010, when a new image transmitted by the *Mars Reconnaissance Orbiter* revealed signs of severe ice damage to the lander's solar panels. Nevertheless, the *Phoenix* spacecraft succeeded in its investigations and exceeded its planned lifetime, providing an important step 'to spur the hope that we can show Mars was once habitable and possibly supported life'.[7] Furthermore, the soil chemistry and minerals at the site revealed it has been in a wet and warm climate over the last few million years, implying that it may meet the criteria as a suitable habitat for future astronauts.

Curses!

In 2011 Russia temporarily returned to the Mars scene with its *Phobos-Grunt* sample-return probe. This daring mission, aimed at reviving exploratory missions to Mars's moons and returning a sample to Earth, ended shortly after lift-off on 8 November when its booster failed to fire while still in low Earth orbit, adding another chapter to the Soviet, and later Russian, 'space disaster series':[8] out of 21 attempts to get to the Red Planet, eighteen were unsuccessful. The loss of *Phobos-Grunt* also meant a swift end to China's first interplanetary probe, *Yinghuo-1*, a tiny cube spacecraft, each side measuring less than a metre and massing only 115 kilograms (255 lb), which tried to hitch a ride to Mars on the Russian probe, and also to the equally small *Living Interplanetary Flight Experiment* funded by the Planetary Society, to investigate the transpermia hypothesis, the idea that a living organism might survive a journey through space to Earth inside a meteorite.

A Roving 'Curiosity'

After about eight months of interplanetary travel, NASA's *Mars Science Laboratory Curiosity* touched down on 5 August 2012 inside Mars's Gale Crater, in which previous missions had detected past evidence of water, and soon began its primary mission to determine whether climatic conditions on the Red Planet were ever suitable to microbial life. The car-sized, six-wheeled rover, about as tall as a basketball player, is the largest and most complex machine ever sent to Mars. The rover's large size (3 metres (10 ft) long, 2.2 metres (7.2 ft) tall and 2.7 metres (8.8 ft) wide) allows it to carry an advanced kit of ten scientific instruments, including seventeen cameras, a laser to vaporize and study small pinpoint spots of rocks at a distance, and a drill to collect powdered rock samples with its 2 metres (6.5 ft) arm. These are the biggest, most advanced instruments ever sent to Mars.

In addition to landing in a place with past evidence of water, *Curiosity* is seeking evidence of organics, the chemical building blocks of life. The history of Martian climate and geology is written in the chemistry and structure of the rocks and soil. *Curiosity* reads this record by analysing powdered samples drilled from rocks. Microbes need three conditions to exist: water (particularly non-acidic) to facilitate chemical reactions, sources of energy and organic (carbon-bearing) material, and enough time for chemistry to occur.[9] *Curiosity* also measures the chemical fingerprints present in different rocks and soils to determine their composition and history, especially their past interactions with water.[10]

Gale Crater is a fascinating Martian landmark that may have been a river or lake bed in ancient times. The crater, an impact scar roughly 150 kilometres (93 mi.) wide just south of the planet's equator, has a key central mountain (Mount Sharp), which rises 5 kilometres (3 mi.) high. Its layered appearance suggests the mountain is a surviving remnant of an extensive series of deposits laid down after a massive impact that excavated Gale more than

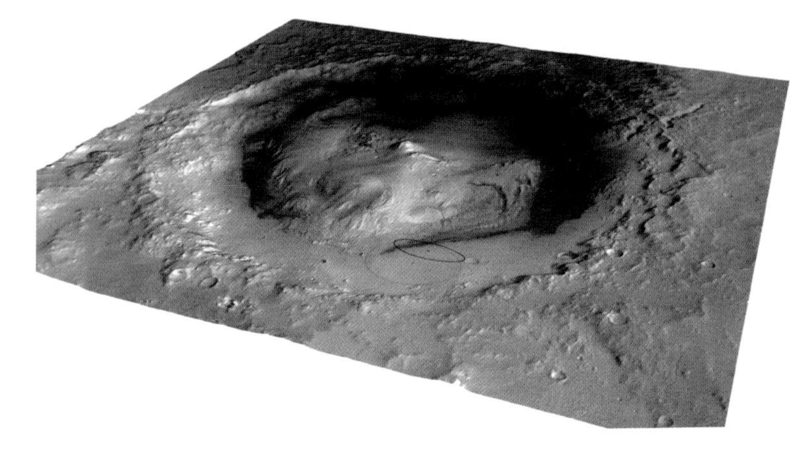

Mars Reconnaissance Orbiter image of Gale Crater, where the *Curiosity* rover is currently exploring. *Curiosity* has managed to detect the presence of tridymite within the crater, hinting at the presence of powerful, explosive volcanism on the ancient Martian surface.

3 billion years ago. The layers offer a history book of sequential chapters recording environmental conditions when each stratum was deposited.

While the rover was fit to climb over knee-high obstacles and travel about 30 metres (100 ft) per hour, the craft's engineering team decided to land the rover far from the mountain on flat-rock terrain and then, from there, work its way across mundane hummocky terrain to what was to be the first stop on its tour, a geological wonderland known as Yellowknife Bay. Only weeks after arriving, *Curiosity* rolled across an ancient stream bed – where some stones were large and round, meaning that water could have transported them over long distances – and hit pay dirt: chemical analysis of powder drilled out of a rock provided the best evidence yet that the Red Planet could have supported living microbes billions of years ago. 'I think this is probably the only definitively habitable environment that we have described and recorded,' said David Blake, a scientist at NASA's Ames Research Center who is the principal investigator for *Curiosity*'s CheMin lab.[11]

The ingredients included nitrogen, oxygen, hydrogen, sulphur, phosphorus and carbon – the elemental ingredients, or building blocks, for life. While this was not an announcement of life itself, the findings were nonetheless encouraging. Analysis of additional drilled samples later lent support to Gale Crater likely hosting a potentially habitable freshwater lake-and-stream system for millions of years in the ancient past. Clay minerals made up at least 20 per cent of the samples, indicating that the water available during the formation of the rock at Yellowknife Bay, billions of years ago, was fresh enough for a human to drink.

Curiosity remained in Yellowknife Bay for seven months before it rapidly rolled towards its primary goal, the base of Mount Sharp. Along the way the craft repeatedly sniffed the air for signs of methane gas, a tell-tale sign of life, at least on Earth. During the journey the rover detected wildly varying amounts of low-level methane, which,

This illustration shows the ways in which methane from the subsurface might find its way to the surface where its uptake and release could produce a large seasonal variation in the atmosphere, as observed by *Curiosity*.

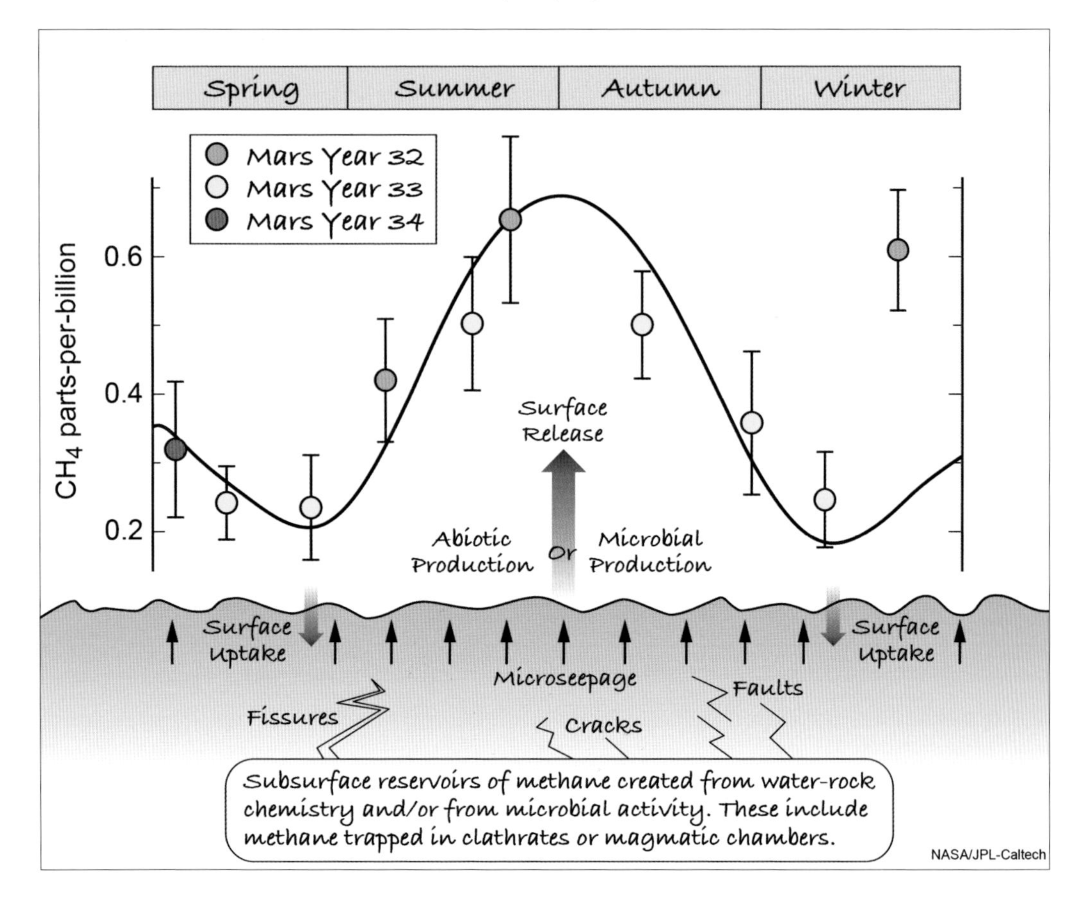

Self-portrait of NASA's *Curiosity* rover on 15 June 2018, as a regional dust storm was about to balloon into a planet-encircling dust event. Note how the airborne dust has turned the sky pink and covered the Martian soil with ochre dust. Fortunately, the rover was largely unaffected by the storm.

on occasion, skyrocketed from mere traces of gas (0.3–0.8 parts per billion) to a tenfold increase (about 7 parts per billion), fuelling debates as to the origin of these methane spikes.

Current explanations run the gamut from periodic meteor showers raining down the gas to ultraviolet radiation breaking up organic material on the ground; and no one has completely ruled out seasonal variability of microbial life in the Martian soil.[12]

After fourteen months of trundling, the rover reached its destination in late 2014, where it began a more ambitious hunt for more complex organic molecules. A major breakthrough for astrobiologists occurred in June 2018 when the rover detected carbon-containing compounds preserved in 3-billion-year-old sediments. Heating of the sediments released an array of organics and volatiles reminiscent of organic-rich sedimentary rock found on Earth. Jennifer Eigenbrode, a biogeochemist at NASA's Goddard Space Flight Center in Greenbelt, Maryland, said that, once again, while these are not evidence of life itself, the discovery presents 'compelling evidence' that Martian life could have existed in the past, adding that 'if it did, we have a chance of finding evidence of it'.[13]

While the carbon molecules and methane variations could be linked, it is unclear whether either are caused by a biological or non-biological process. Nevertheless, when the two are combined, these discoveries increase the chances not only that microscopic organisms may have once thrived on Mars, but that they still exist. *Something* appears to be performing ongoing reactions deep below the surface of Mars, and Eigenbrode says those reactions 'could potentially be related to liquid water or life'.

As of this writing, *Curiosity* has spent more than 2,574 Earth days on Mars, placing it second behind the *Opportunity* rover (5,170 days) in terms of total time spent on the Red Planet. It has also travelled more than 20 kilometres (12.5 mi.), studied more than 180 metres (590 ft) of vertical rock, survived a two-month-long global dust storm, and is preparing to roll on with no signs of stopping. In June 2019, *Curiosity* detected the largest amount of methane ever measured during the mission – about 21 parts per billion units by volume (ppbv). One ppbv means that if you take a volume of air on Mars, one billionth of the volume of air is methane. While microbial life is an important source of methane on Earth, the gas can also be created through interactions between rocks and water. Only days after NASA announced the official end to the *Opportunity* rover, *Curiosity* entered 'safe mode' due to an unexplained glitch, which may have also been caused by the dust storm that led to *Opportunity*'s demise. NASA's JPL called the glitch a 'hiccup' at the time, but now it seems everything is back to normal.

As it rolls, the rover continues to sample the soil and sniff the air, collecting data on solar radiation and weather that will help us learn more about whether Mars could be habitable for humans on

This image shows part of the *MAVEN* spacecraft – a cube roughly 2 m flanked by solar panels that together span approx. 6 m – and the limb of Mars in the background. The image shows the magnetometer and Sun sensor at the end of the solar panel. The dark spot at the top of the image is Mars's Olympus Mons volcano.

a local level, a role magnified on a global scale with the arrival of the orbiting MAVEN spacecraft.

MAVEN *Investigates the Martian Atmosphere for Future Astronauts*

On 21 September 2014, ten days after *Curiosity* reached Mount Sharp, NASA's *Mars Atmosphere and Volatile EvolutioN* (MAVEN) spacecraft achieved Mars orbit and began sniffing the Martian upper atmosphere for its constituent gases, diving to within 160 kilometres (100 mi.) of the surface at times as it seeks out new information on the planet's climate and potential for habitability.

Astronomers have long believed that Mars was once a wetter and warmer world with conditions friendly to life as we know it. The Mars of today, however, is but a shadow of its former self. As discovered in the pre-spacecraft era, the present Martian atmosphere is woefully thin, less than 1 per cent the atmospheric volume of Earth's, which is too cold and tenuous for liquid water to persist on the surface, where it would easily boil away. Its surface is now anything but friendly to life, being exposed to intense radiation

This artist's concept depicts the early Martian environment (left), believed to have contained water and have had a thicker atmosphere environment than Mars today (right).

and susceptible to wild swings in temperature by as much as 14°C (57°F) in an odd, twice-a-day pattern; as MRO discovered, variation in solar heating between day and night drive these semi-diurnal atmospheric tides.[14] Given the data collected by the *Curiosity* rover, which paints a picture of an ancient Earth-like Mars – one with a climate warm enough to have sustained water at the surface long enough in the form of river deltas and lakes for life to emerge and evolve – one big question remained: where has Mars's former habitable atmosphere gone?

With Mars being so close to the Asteroid Belt, initial thinking came up with a dramatic scenario that there were numerous planetary impacts by comets and asteroids early in its history. A massive impact could have caused critical damage to Mars's internal dynamo, 'switching off' its global magnetic field. The impact could have also blasted huge quantities of atmospheric gases into space. With no magnetic field to act as a force field against the Sun's continuous barrage of energetic particles, over time the remaining Martian atmosphere was literally stripped into space.

MAVEN's studies since its arrival have revealed that solar wind and radiation are indeed the culprits responsible for stripping away most of the Martian atmosphere and transforming Mars from a planet that could have supported life billions of years ago into a frigid desert world. The solar wind is a steady stream of particles, mostly protons and electrons, emitted by the Sun. The early Sun had far more intense ultraviolet radiation and solar wind, so atmospheric loss by these processes was likely much greater in Mars's history. Since MAVEN arrived at Mars in 2014, its investigations have found that this atmosphere may have been stripped away by a torrent of solar wind over several hundred million years, between 3.5 and 4 billion years ago.[15]

In their study MAVEN researchers focused on two types of argon: argon-36, a lighter variant of the gas found in abundance in the upper atmosphere, and argon-38, a heavier form of the gas found at lower altitudes. Argon is a 'noble' gas that doesn't react chemically with other elements, so once it is in the atmosphere it should stay there, unless removed by force. And this, the researchers found, is what happens to argon in Mars's atmosphere, especially the lighter argon-36 in the upper atmosphere.

Unlike Earth, Mars does not have a strong magnetic field to protect it against the Sun's radiation and solar wind, a thin stream of electrically conducting gas constantly blowing out from the surface of the Sun. This stream contains electrically charged particles (ions), which impact Mars at high speed and physically knocks atmospheric gas into space in a process called 'sputtering'.

By studying the ratios of the lighter vs heavier variants of argon (because it can be removed only by sputtering), the MAVEN team found that about 65 per cent of Mars's argon has disappeared from the Red Planet since it formed. Once they determined the amount of argon lost by sputtering, they used this information to determine the sputtering loss of other atoms and molecules, including carbon dioxide (CO_2, an efficient greenhouse gas that can retain heat and warm the planet), and determined that the majority of the planet's CO_2 was also lost to space by sputtering. As other processes can scrub CO_2, the research provides the minimum amount of CO_2 that has been lost to space.[16] Working back, they modelled the ancient atmosphere of Mars, finding that billions of years ago it was very much like Earth's.

In an ironic twist, Stephen Mojzsis of the geological sciences department of the University of Colorado Boulder and colleagues believe that the massive impacts that may have removed the planet's magnetic field may have also enhanced climate conditions enough to make the planet more conducive to life – before the effects of sputtering took effect.[17] In addition to producing hydrothermal regions in portions of Mars's fractured and melted crust, a massive impact could have temporarily increased the planet's atmospheric pressure, periodically heating Mars up enough to 're-start' a dormant water cycle. The researchers say this would have helped life thrive, if life was present, but according to Mojzsis, 'up to now we have no convincing evidence life ever existed there, so we don't know if early Mars was a crucible of life or a haven for life.'[18]

Three days after MAVEN achieved Mars orbit, the Indian Space Research Organisation launched its first interplanetary mission to Mars, the *Mars Orbiter Mission* (MOM; formerly *Mangalyaan*) followed suit, making India the first country to enter the Red Planet's orbit on its first attempt. Furthermore, the spacecraft's budget of 450 crore rupees (U.S.$73 million) makes it the least expensive interplanetary mission to date to reach Mars.[19] MOM is also the only orbiting spacecraft that can image the full disc of Mars in one frame and also image the far side of the planet's moon Deimos.

Its main six-month mission was to test key technologies for interplanetary exploration and to use its five science instruments to collect data of Mars's morphology, atmospheric processes, surface temperature, surface geology and how gases are escaping the planet's atmosphere – which complements NASA's MAVEN mission. MOM has surpassed its six-month mission and, as of 24 September 2019, completed five years of orbiting Mars. It

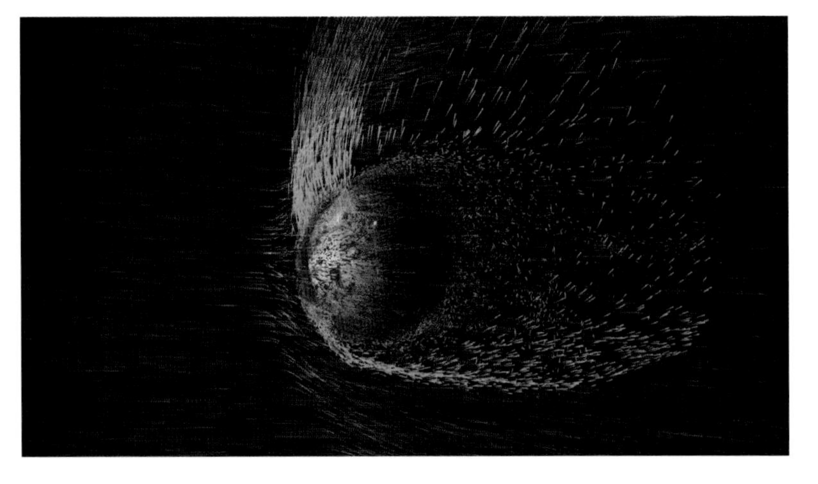

Unlike Earth, Mars lacks a strong magnetic field that deflects the stream of charged particles continuously blowing off the Sun. Instead, the solar wind crashes into the planet's upper atmosphere, stripping away atoms (depicted here in a simulation).

Side-by-side images show how dust enveloped the Red Planet between 28 May 2018 and 1 July 2018, courtesy of the Mars Colour Imager wide-angle camera on board NASA's *Mars Reconnaissance Orbiter*.

continues to collect data for its primary science mission. MOM is built with full autonomy to take care of itself for long periods without any ground intervention. As of this writing, MOM has returned thousands of images totalling two terabytes, including close distance images of Phobos and Deimos.

India's MOM arrived in time to join the armada of international spacecraft orbiting the planet in observing a key atmospheric event: the global dust storm of 2018. All dust events, regardless of size, help shape the Martian surface. Studying their physics is critical to understanding the ancient and modern Martian climate, according to Rich Zurek, chief scientist for the Mars Program Office at the Jet Propulsion Laboratory in Pasadena: 'Each observation of these large storms brings us closer to being able to model these events – and maybe, someday, being able to forecast them. That would be like forecasting El Niño events on Earth, or the severity of upcoming hurricane seasons.'[20]

The first sign of storm activity occurred in April 2018 when the ESA's *Mars Express* orbiter captured images of a local dust storm with a towering cloud front, located close to the northern polar region of the Red Planet.[21] This storm occurred just weeks before the start of the global storm that intensified into one of the most severe storms ever observed on Mars. Called the 'perfect storm' for science, it

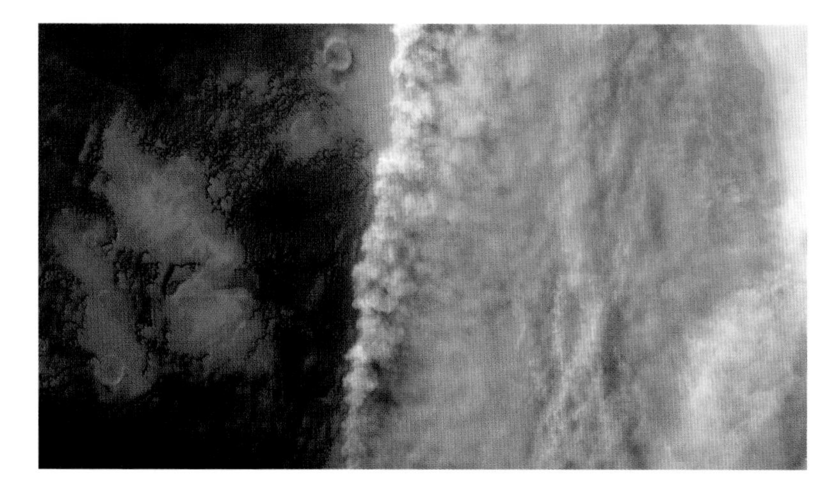

On 3 April 2018 the high-resolution stereo camera on board ESA's *Mars Express* orbiter captured this impressive upwelling front of dust clouds – visible in the right half of the frame – near the north polar ice cap of Mars. The event turned out to be a harbinger of a historic dust storm on Mars.

occurred with a historic number of robotic Mars explorers on and orbiting the planet.

Mars's thin atmosphere makes these storms vastly different from anything encountered on Earth. Although they are not strong enough to topple a spacecraft, as depicted in the film *The Martian* (2015), they can sandblast dust particles into the atmosphere. Indeed, MAVEN confirmed *Mars Reconnaissance Orbiter* data that global dust storms on Mars transport not only dust into the middle atmosphere (roughly 50–100 km (30–60 mi.) high), but lesser amounts of water vapour, driving a seasonal pattern that hastens the loss of hydrogen from the top of Mars's atmosphere. Once the water vapour is higher aloft, high-energy ultraviolet radiation from the solar wind can split the water molecules (H_2O) apart into two single hydrogen atoms and one atom of oxygen. Several processes at work in Mars's upper atmosphere, such as turbulence when nearest the Sun, may then act on the hydrogen, leading to its escape into space.[22] Jim Watzin, director of NASA's Mars Exploration Program at the agency's headquarters in Washington, DC, said, 'Each offers a unique look at how dust storms form and behave – knowledge that will be essential for future robotic and human missions.'[23]

'ExoMars *Trace Gas Orbiter*' Arrives at Mars

In April 2018, as the ESA's *Mars Express* orbiter mission was capturing images of dust welling up on the surface of Mars, the European Space Agency/Roscosmos *ExoMars Trace Gas Orbiter* mission, launched in 2016 and fitted with instruments from both Europe and Russia, reached its final orbit after a year of aerobraking and began its search for gases that may be linked to active geological or biological activity on the Red Planet.

The primary goal of the craft's science mission is to provide the most detailed inventory of Martian atmospheric gases to date, including one of trace gases that make up less than 1 per cent of the total volume of the planet's atmosphere, such as methane and other gases that could be signatures of active biological or geological activity. Methane on Mars is expected to have a rather short lifetime of about four hundred years, so any detections imply it must have been produced or released relatively recently. *ExoMars* will build up a picture over time of the methane distribution, to understand

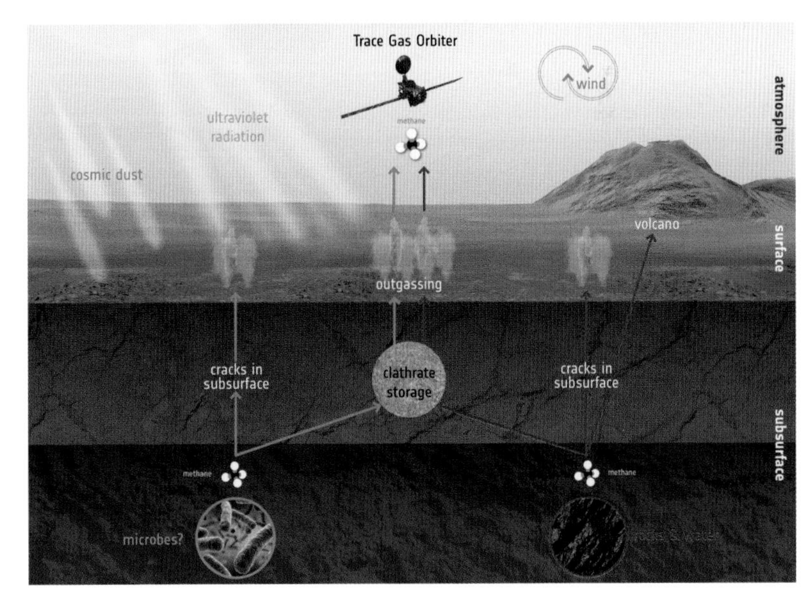

The *ExoMars Trace Gas Orbiter* is set to analyse the Martian atmosphere. This graphic depicts some of the possible ways methane might be added or removed from the atmosphere. One exciting possibility is that methane is generated by microbes. If buried underground, this gas could be stored in lattice-structured ice formations known as clathrates, and released to the atmosphere at a much later time.

geographic and seasonal distributions, and eventually home in on areas where it might be originating.[24]

Initial results from the craft's study of the Martian atmosphere during the 2018 global dust storm, presented in April 2019 at a European Press meeting in Vienna, revealed an enhancement of water vapour that happened remarkably quickly, over just a few days during the onset of the storm, indicating a swift reaction of the atmosphere to the dust storm. As reported by the European Space Agency, the observations are consistent with global circulation models, whereby dust absorbs the Sun's radiation, heating the surrounding gas and causing it to expand, in turn redistributing other ingredients – like water – over a wider vertical range. A higher temperature contrast between equatorial and polar regions is also set up, strengthening atmospheric circulation. At the same time, thanks to the higher temperatures, fewer water-ice clouds form – normally they would confine water vapour to lower altitudes.

Aside from microbial activity, methane can also be generated by reactions between water and olivine-rich rocks, perhaps in combination with warmer, volcanic environments. Again, this could

Visualization of the *ExoMars Trace Gas Orbiter* aerobraking at Mars. With aerobraking, the spacecraft's solar array experiences tiny amounts of drag owing to the wisps of Martian atmosphere at very high altitudes, which slows the craft and lowers its orbit.

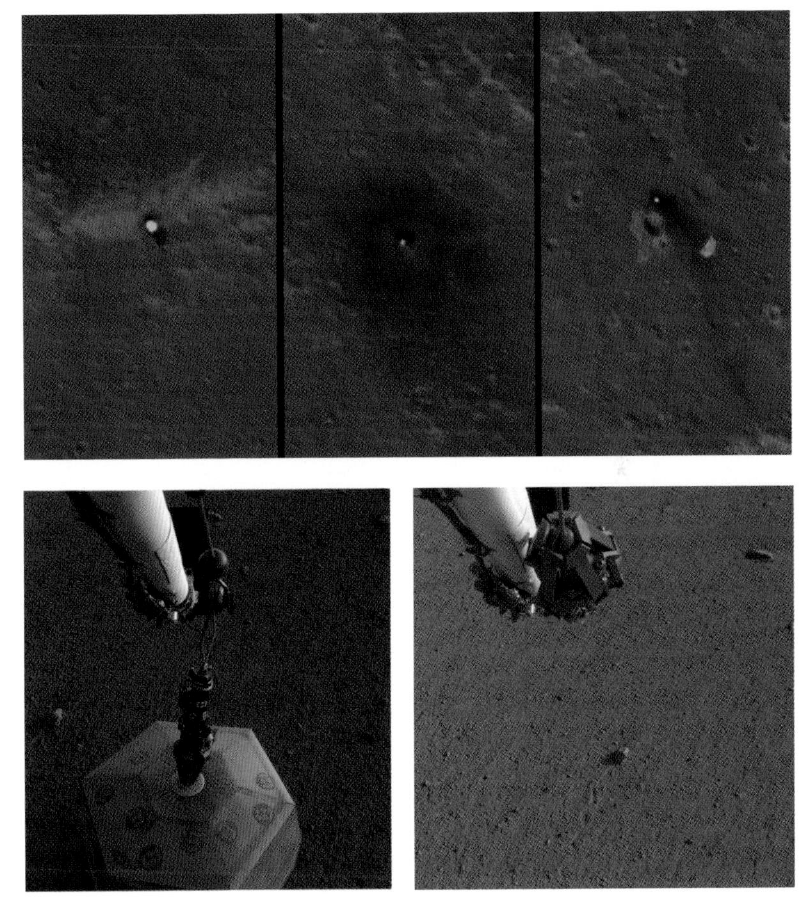

On 26 November 2018, NASA's *InSight* spacecraft successfully touched down on Mars. The *InSight* spacecraft, its heat shield and its parachute were imaged on 6 and 11 December 2018 by the HiRISE camera on board NASA's *Mars Reconnaissance Orbiter*.

Left: On 19 December 2018, *InSight* lander deployed its first instrument – a seismometer – onto the surface of Mars, allowing scientists to peer into the Martian interior by studying ground motion (also known as marsquakes).

Right: The rock in the centre of this image was tossed about 1 m (3 ft) by NASA's *InSight* spacecraft as it touched down on Mars on 26 November 2018. The rock, which is a little bigger than a golf ball, was nicknamed in August 2019 'Rolling Stones Rock' in honour of The Rolling Stones.

be stored underground in icy cages. Such data will support part of ESA's *Mars Express* orbiter's mission to use ground-penetrating radar to discover large water reservoirs of water scattered across its polar regions (which it did in July 2018, finding a pond of liquid water buried under layers of ice and dust in the south polar region of Mars).[25] Finding them and figuring out how to access their bounty could be critical for potential human explorers travelling to Mars in the future.

ExoMars has also mapped the distribution of hydrogen in the uppermost metre of the planet's surface, which is important as hydrogen is one of the constituents of the water molecule. The initial maps presented at the April 2019 European Press meeting – based on just a few months' data – have a higher resolution than those produced from sixteen years of data from NASA's *Mars Odyssey*. Aside from the obviously water-rich permafrost of the polar regions, the new map provides more refined details of localized 'wet' and 'dry' regions. It also highlights water-rich materials in equatorial regions that may signify the presence of water-rich permafrost in present times, or the former locations of the planet's poles in the past.

Radiation and Habitation

According to Jordanka Semkova of the Bulgarian Academy of Sciences and lead scientist of the Liulin-MO instrument aboard *ExoMars*:

> One of the basic factors in planning and designing a long-duration crewed mission to Mars is consideration of the radiation risk. Radiation doses accumulated by astronauts in interplanetary space would be several hundred times larger than the doses accumulated by humans over the same time period on Earth, and several times larger than the doses of astronauts and cosmonauts working on the International Space Station.[26]

Once beyond the protective cage of the Earth's strong magnetic field and thick atmosphere, astronauts are exposed to the unceasing bombardment of galactic cosmic rays and other radiation that can lead to radiation sickness, an increased lifetime risk for cancer, central nervous system effects and degenerative diseases. *ExoMars*'s radiation measurements have revealed that astronauts on a mission to Mars would be exposed to at least 60 per cent of the total radiation dose limit recommended for their career during the journey itself to and from the Red Planet.[27]

Curiosity also has a Radiation Assessment Detector (RAD), one of the first instruments sent to Mars specifically to prepare for future human exploration. The instrument is the size of a six-pack of fizzy drinks or soda and measures and identifies all high-energy radiation on the Martian surface that could pose a risk to humans, including not only direct radiation from space (such as gamma-rays), but secondary radiation produced by the interaction of space radiation with the Martian atmosphere and surface rocks and soils. Additionally, an ultraviolet sensor stuck on *Curiosity*'s deck tracks radiation continuously.

Measurements made by RAD during the flight to Mars and on the planet's surface allowed NASA scientists to estimate the amount of radiation astronauts would be exposed to on an expedition to Mars, leading them to determine that the levels were manageable for a crewed Mars mission in the future.[28] The RAD instruments suggest that a mission with 180 days flying to Mars, five hundred days on the surface and 180 days heading back to Earth would create a dose of 1.01 sieverts (a measurement unit of radiation exposure to biological tissue).

The total lifetime limit for European Space Agency astronauts is 1 sievert, which is associated with a 5 per cent increase in fatal cancer risk over a person's lifetime. NASA currently caps the risk limit at 3 per cent, but adjustments could be made to accommodate trips outside of low Earth orbit. 'It's certainly a manageable number,' according to Don Hassler, a programme director at the Southwest Research Institute in Boulder, Colorado, and principal investigator of the RAD investigation.[29] RAD measurements inside shielding provided by the spacecraft show that such a mission would result in a radiation exposure of about 1 sievert, with roughly equal contributions from the three stages of the expedition. A key factor in achieving travel to Mars is to have means in place to protect astronauts from space radiation. Finding ways to limit the harmful effects are increasingly important as longer missions further from Earth are being considered.

More than a century has passed since Percival Lowell postulated a race of intelligent beings on Mars, who learned how to channel water from the planet's polar ice caps to its arid surface in order to sustain their race on this dying world. His conjectures were, in time, scoffed at by professionals who soon began to use data from scientific instruments to counter Lowell's fancies, ironically looking upon Mars as a dead and helpless planet. Continuing the irony, humans have since come to visualize intelligent beings on Mars – ones who will be seeking ways to channel water to their colonies in order to help them survive the ill effects of this inclement planet. The fact is that we are these foreseeable beings who have come to long for Mars as an alternate home for humanity. It almost seems as if Lowell had foreseen our future whenever he spied Mars through his telescope and fired his imagination.

In 2016 the British theoretical physicist Stephen Hawkings expounded that humanity needs to find an alternative home in the next two hundred to five hundred years if it is to survive extinction in 1,000 years . . . or perhaps even in the next century, noting that the Earth is under so many threats that he finds it difficult to be positive. 'Spreading out may be the only thing that saves us from ourselves,' he told the *Washington Post* in June 2017.[30] 'We are running out of space, and the only place we can go to are other worlds,' he said via video link to the audience gathered at the 2017 Starmus Festival in Trondheim, Norway, 'I am convinced that humans need to leave Earth.'[31]

Terraforming – the process of creating an Earth-like or habitable environment on another planet via human intervention – Mars is out . . . unless you want to wait 100,000 years for the process to kick in, making the Red Planet's atmosphere breathable. Proponents of terraforming Mars propose releasing greenhouse gases from a variety of sources on the Red Planet to thicken the atmosphere and increase the temperature to the point where liquid water is stable on the surface. But researchers who have analysed the abundance of carbon-bearing minerals and the occurrence of CO_2 in polar ice, using data from the *Mars Reconnaissance Orbiter* and *Mars Odyssey* spacecraft, as well as data on the loss of the Martian atmosphere to space from MAVEN, suggest that there is not enough CO_2 left on Mars to provide significant greenhouse warming were the gas to be put into the atmosphere. In addition, most of the CO_2 gas is not accessible and could not be readily mobilized. As a result, terraforming Mars is not possible using present-day technology.[32]

Hawking instead suggests going first to the Moon and building a base there, before repeating the base-building process on Mars. Hawking's thinking is aligned with current ESA and NASA long-term plans to colonize Mars: Jan Wörner, Director General of ESA, said in 2016 that a 'moon village' would take twenty years to plan and construct, and NASA's long-term plans include sending humans to Mars by the 2030s. The question is, what will it take to succeed?

Can we colonize Mars?

SEVEN

Mars: Our Home Away from Home?

Can Mars, a mysterious world, desolate and extreme, serve as a safe haven for future space explorers? Harsh conditions abound on Mars. The planet's atmosphere is 95 per cent carbon dioxide (0.13 per cent oxygen) and one hundred times thinner than Earth's at sea level. Its gravity is 37 per cent that of Earth's. Average temperatures hover around –63°C (–81°F), about as cold as inhabited Oymyakon, Siberia, which logged a record low of –69°C (–92°F) in 2013.[1] Crushed volcanic rock and dust (*regolith*), which lacks anything organic like soil, covers the planet's frozen surface, which is very similar to the weathered basaltic materials found in the high-altitude landscapes of Hawaii's Maunakea and Maunaloa volcanoes.[2] And global dust storms threaten to sweep across the planet around the time of southern spring or summer. As Claude A. Piantadosi writes in discussing human survival in extreme environments, 'NASA's interest in [resource utilization *in situ*] for Mars missions implies that the problem of extended life support is, indeed, formidable.'[3]

Nevertheless, the Red Planet continues to capture the imagination of visionaries who foresee a future for humanity on Mars. The robotic missions we have sent thus far to the Red Planet have brought us closer to interpreting the world as it is – an airless body (replete with geology and devoid of conspicuous biology). But as the nineteenth-century British artist and visionary John

Ruskin (1819–1900) believed, close observation is only a starting point for the true force: imagination.[4]

NASA's latest plan is not a point-and-shoot mission to Mars, but a tiered learn-as-we-go process that culminates with sending the first humans to the planet in the 2030s.[5] The plan has three principal stages: Earth Reliant, Proving Ground and Earth Independent. Each step will have increasing challenges, as the explorers move progressively farther from Earth.

Stage 1 (Earth Reliant) is already underway with the research conducted aboard the International Space Station (ISS), where astronauts learn how to live and work in space. Earth Reliant will help uncover the cumulative effects of exposure to space on the human body and perform research that will enable deep space, long-duration missions, among other things.

In Stage 2 (Proving Ground), NASA will conduct complex operations in laboratories primarily in cislunar space (the volume of space around the Moon) in the 2020s to advance and validate capabilities required for humans to live and work at distances much farther away from our home planet than where the ISS and other space stations operate. Proving Ground may also feature multiple staging orbits for future deep space missions, which will allow crews to return to Earth in a matter of days, if necessary.

Stage 3 (Earth Independent) activities build on what will have been learnt in Stages 1 and 2 to enable human missions, first perhaps to low Mars orbit (or one of the Martian moons) and eventually to the surface of the planet. NASA expects future Mars missions will represent a collaborative effort with partners who wish to seek the potential for sustainable life beyond Earth. Science missions are already in the pioneering stages of the Earth Independent phase. In September 2017 Lockheed Martin revealed its proposal, suggesting NASA work with its international partners and private industry to set up a space station (Mars Base Camp)

in Mars orbit by 2028, where astronauts will conduct exploratory science to confirm where best to land humans on the surface in the 2030s.[6]

Martian landscape (left) with *Sojourner* rover, and Maunakea summit terrain (right).

The Challenge of Mars

> Only those who will risk going too far
> can possibly find out how far one can go.[7]

A piloted journey to Mars will be a frontier endeavour in a climate of risk, as astronauts will experience radiation dangers, plus physical and mental health conditions never faced before. But scientists are optimistic that they will find solutions to the challenges ahead.

Stage 1 research is already blazing the trail for safer deep-space journeys. Radiation studies are foremost, as a human mission to Mars means sending astronauts into interplanetary space for a minimum of a year, where they will be exposed to the harsh radiation environment of space. Thus astronauts need to be protected from two major sources of radiation: solar radiation and galactic cosmic radiation.

The Sun not only releases a steady stream of solar particles into space, but erupts violently with occasional flares and coronal mass ejections that send out an expanding wave of high-energy particles (almost all protons). Fortunately, the proton energy is low enough that they can almost all be physically shielded

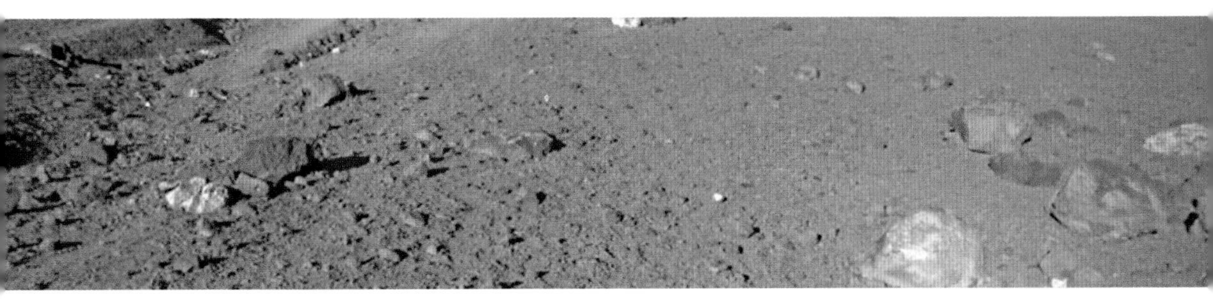

by the structure of the spacecraft. NASA currently operates a fleet of spacecraft studying the Sun and the space environment throughout the solar system, giving us a better understanding of the origin of solar eruptions and what effects these events have on the overall space radiation environment.

Galactic cosmic rays are much more hazardous. This high-energy radiation shoots into our solar system from other stars in the Milky Way (or even other galaxies), and can pass right through the skin, damaging cells or DNA along the way that can mean an increased risk for cancer later in life or, at its worst, acute radiation sickness. NASA scientists have already started working on solutions, including using new materials that would shield the spacecraft and astronauts from this harmful radiation.[8] One possibility includes hydrogenated boron nitride nanotubes (BNNTs), made of carbon, boron and nitrogen, with hydrogen interspersed throughout the empty spaces between the tubes. Researchers have successfully made yarn out of BNNTs, so it is flexible enough to be woven into the fabric of spacesuits, providing astronauts with significant radiation protection even while they are performing spacewalks in transit or out on the harsh Martian surface. Though hydrogenated BNNTs are still in development and testing, they have the potential to be one of the key structural and shielding materials in spacecraft, habitats, vehicles and spacesuits that will be used on Mars.

Of special interest in Stage I research is NASA's Twins Study. The first of its kind, the study compared molecular profiles from before, during and after astronaut Scott Kelly's eleven-month mission (2013–14) aboard the ISS with those of his identical twin brother (retired astronaut Mark) on Earth. Preliminary results have shown that space travel causes an explosion in the way genes are turned on and off.[9] When a gene turns on, information stored in our DNA is converted into instructions for making proteins, which, in turn, builds cells; when they turn off so too does cell production. By toggling this switch, gene expression allows cells to respond to changes in environment, such as weightlessness. It also determines the body's ability to deal with stress and the development of diseases, such as cancer or metabolic disorders.

When Scott Kelly left Earth he shared an identical DNA profile with his brother Mark, but not when he returned. Chris Mason, Principal Investigator of the study, says that the changes began as soon as Scott went into space, causing thousands upon thousands of genes to turn on and off, and some of the activity persisted temporarily upon his return to Earth (when he measured 5.08 cm (2 in.) taller). Mason explains that 'there are over 50,000 genes in the human genome, and when floating in zero gravity, the body is trying to manage that situation in new ways.'[10] So the new result might mean spending more time in space can lead to unprecedented changes in cell function: the findings 'set the bedrock for understanding molecular risks for space travel as well as ways to potentially protect and fix those genetic changes.'[11]

It has long been known that space flight wreaks havoc on the body. In space, muscles atrophy because they no longer need to fight gravity, requiring astronauts to exercise up to three hours a day to maintain muscle mass, cardiovascular fitness and bone density. In micro-gravity an imbalance occurs in the way cells form new bone tissue and destroy old bone tissue, with more removal occurring than growth, causing calcium (and other minerals needed to construct bone) to leach out in urine, weakening the bones.[12] Tests have shown that rate of decay to be constant, though it may plateau during longer stays. Astronauts also experience eye problems, including abrasions from floating specks in space labs, as well as an irreversible deterioration caused by a fluid shift to the head that bulges into the back of the eyeball and changes the shape of the lens.

Other side effects include a weakened immune system (colds and flus can spread like wildfire in a contained environment in space), and it appears bacteria are more dangerous in that weightless environment. Then there's cognitive sluggishness – not to mention stress. Marc Jurblum, a member of the Australasian Society of Aerospace Medicine's Space Life Sciences Committee, questions what months and months of living in an unchanging capsule habitat with only blackness outside the small window will do to an astronaut's mind.[13] Working with research groups looking at how to maintain mental health in extreme environments, he is trying interventions, such as meditation, positive impact pictures of nature and virtual reality, to see if they have a healing effect on space travellers. Already astronauts use Mission Control as a sounding board to alleviate anger and stress, so they can continue to act quickly and react as a team.

John Bradford, president and chief operating officer of SpaceWorks Enterprises in Atlanta, Georgia, believes one way to ease the journey's physical and psychological burden is to place the astronauts in a state of hibernation known as 'therapeutic hypothermia'.[14] Lowering astronauts' body temperatures by about 5°C (40°F) would induce a 'hypothermic stasis' that cuts crew members' metabolic rates by 50 to 70 per cent. This would allow astronauts to more or less sleep their way to Mars in a rotating habitat that creates artificial gravity, which is essential to long-duration missions as it will counteract the deleterious effects of skeletal deterioration, muscle atrophy, weakened cardiovascular systems, severe vision problems and more.

Andy Weir, author of *The Martian*, wrote an opinion piece in 2014 on what needs to be resolved:

> The physics are simple: Make a ship that can withstand one g of force (equivalent to the force of gravity on Earth's surface), then start the ship spinning such that the centripetal force is one g at the edge. That's it. No fancy 'Star Trek' technology needs to be invented. The ship just needs to spin. And once it's spinning in the vacuum of space, it requires no additional energy or maintenance to continue.[15]

Weir adds that designing a station with artificial gravity would undoubtedly be a daunting task: 'But the cost of *not* having artificial gravity is proving to be massive, and it is an absolute requirement for manned exploration of our solar system to develop this technology.'

During the journey envisioned by Bradford, astronauts would be fed intravenously and catheterized. The plan will not only save on food, water and oxygen, it will minimize the physical, psychological and social challenges that the crew would otherwise have to face. While challenges remain (more research especially needs to be done on the time needed to 'wake up' the astronauts), Bradford thinks it is all manageable.

How Will We Get to Mars?

At the core of Stage 2 (Proving Ground) is NASA's Space Launch System (SLS), the world's most powerful heavy-lift rocket designed to send six astronauts in the agency's *Orion* spacecraft (described below) into cislunar orbit, then on to Mars.[16] It is also designed to send critical labs, habitats and supplies to Mars, as the super booster offers an unprecedented lift capacity (130 tonnes), as well as the energy to speed missions through space to Mars. The major components of the architecture will be launched separately with some pre-positioned in Mars orbit ahead of time. Others will be assembled in cislunar space.

The Space Launch System is made of three main sections. (1) The core stage (long orange cylinder), which stores 730,000 gallons of super-cooled liquid hydrogen and liquid oxygen that will fuel the system's RS-25 engines; it is bracketed by two (white) solid rocket boosters. (2) The Launch Vehicle Adaptor stage (tapered orange section), which stores the Interim Cryogenic Propulsion Stage. (3) The Orion Multipurpose Crew Vehicle (top) that includes the crew module (for up to six astronauts) and its Launch Abort System that jettisons away after Orion reaches orbit.

Orion is the first exploration vehicle in a generation designed to carry humans into deep space, provide emergency abort capability, sustain the astronauts during their missions and provide safe re-entry to Earth. The *Orion* crew module, which holds six astronauts, will launch on top of the SLS and serves as the heart of the *Mars Base Camp* interplanetary ship. *Orion*'s first crewless flight test on 5 December 2014 was a success. During Exploration Mission-1 (EM-1) *Orion* will travel to the Moon and back in about 26 days. NASA hopes to have *Orion* ready for a crewed flight to the Moon by 2024; NASA has dubbed this mission back to the Moon Artemis, after Apollo's twin sister from Greek mythology. As of October 2019, plans remain for the first uncrewed lunar flyby mission (Artemis-1) to occur sometime between 2020–21. Meanwhile, NASA has partnered with thirteen U.S. companies developing new technologies that possibly could be used by its Artemis programme.

Mars Landing

In September 2017 Lockheed Martin unveiled its single-stage, reusable Mars lander.[17] Called the *Mars Ascent/Descent Vehicle* (MADV), it would attach to the orbiting *Mars Base Camp*, and travel to and from the Martian surface via supersonic retropropulsion, which uses rocket engines to slow the lander from supersonic speeds during its descent. Each surface mission could last two weeks with up to four astronauts, and return to the orbiting *Mars Base Camp* without surface refuelling or leaving assets behind. The lander command deck would use *Orion* avionics and systems powered by engines using liquid hydrogen/liquid oxygen propellant, both of which will be generated from water. Initially the water would come from Earth, but later it could be mined from ice found on asteroids, the cold shadowed craters at the Moon's pole – or from Mars itself.

The more we study Mars, the more water is discovered: in 2002 NASA announced its *Mars Odyssey* spacecraft found enough water ice just beneath the surface to fill Lake Michigan twice over;[18] in 2008, analysis of soil from the NASA *Phoenix* mission's landing site near

A NASA visualization of a landing of the Lockheed Martin *Mars Ascent/Decent Vehicle*.

NASA visualization of a kilopower station on Mars, with four small nuclear reactors to create fission power.

the Martian north pole revealed frozen water;[19] and in January 2018 data from two spacecraft orbiting Mars found cliffs composed mainly of water at eight locations, extending from just below the surface to a depth of 100 metres (330 ft) or more.[20] Human missions to Mars would likely rely on mining water from these known environments. They would then drink the water or break it down into hydrogen and oxygen, which could then be used to make breathable air or methane for rocket fuel.

Researchers at NASA's Los Alamos National Laboratory and the Department of Energy have already successfully tested another source of energy: kilopower.[21] A collection of four or five small and compact nuclear reactors (delivered by a single lander and properly shielded) could generate up to 50 kilowatts. This would be enough to run ten average households on Earth continuously for at least ten years; astronauts on Mars would need only 4 kilowatts to establish a long-term outpost on the planet's surface.[22] That amount of energy could not only provide the necessary lighting, water and oxygen, but be used to produce enough fuel for return journeys to Earth.

Home Away from Home

The habitats our first settler astronauts on Mars will use are still on the drawing board. Whatever the design, it will have to provide the basic needs for long-duration stays: air, water, food and shelter from the frigid, low-pressure, high-radiation environment outside. Of the multitude of ideas that abound, one structure seems to be taking hold: an inflatable Mars Ice Home.[23] Proposed by NASA and designed by Clouds Architecture Office (Clouds AO), Space Exploration Architecture (SEArch) and NASA's Langley Research Center, the structure was inspired by the Mars Ice Home concept that won first prize in NASA's 2015 design challenge.[24]

The Mars Ice Home is a large inflatable torus (like an inner tube) capped by a shell of water ice. Simple robots sent on a single mission can deploy the lightweight habitat, which will incorporate materials extracted from Mars; by extracting 1 cubic metre of water per day from the Martian substrate, the Ice Home design could be completed in just over a year and be ready before the crew arrives. An auxiliary

NASA visualization of the Mars Ice Home concept.

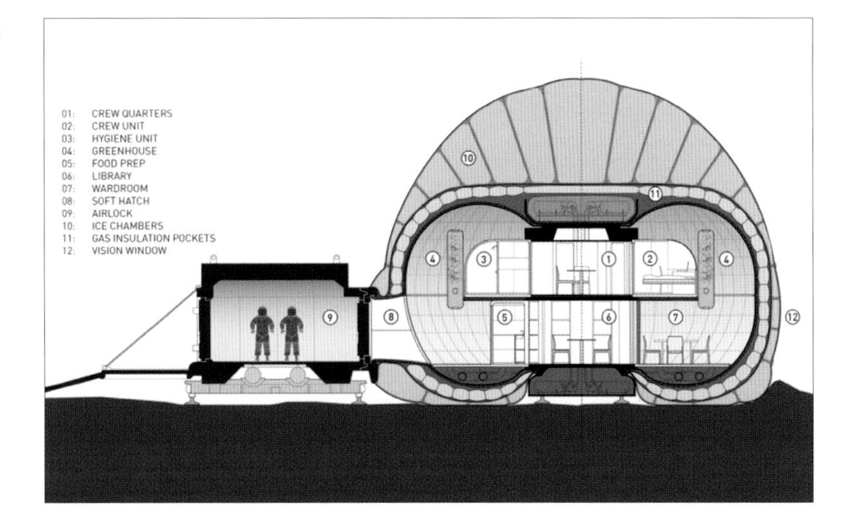

Cutaway view of the Mars Ice Home concept.

ice dome can be used as a storage tank for water, which could potentially be converted later into rocket fuel for the Mars Ascent Vehicle and the return to Earth.

One critical benefit of the Ice Home is that the hydrogen in water ice, which is of similar size to galactic cosmic ray particles, will effectively block and thus shield astronauts on Mars from such radiation exposure; thus the Mars Ice Home seconds as a kind of radiation storm shelter. The thickest part of the dome, which will be above the crew quarters, will also allow sunlight through, helping to illuminate the dome's interior. Managing temperatures inside the Ice Home can be achieved by using a layer of carbon dioxide gas, which can also be extracted from Mars, as an insulator between the living space and ice dome.

NASA announced in 2017 that it would be transporting the Mars Ice Home's wall assembly to the ISS as part of the MISSE-11 mission, which eventually launched in April 2019. The structure will there undergo tests for a year, while ISS astronauts will mount other Ice Home materials to the exterior of the space station to see how they endure the harsh space environment.[25] These materials will then return to Earth for analysis.

What's On the Menu?

Astronauts aboard the ISS are well fed. They eat three meals a day with a balanced supply of vitamins and minerals, and they can hardly complain about the menu as they have some two hundred types of foods to choose from, including fruit, chicken, beef, seafood, sweets and brownies, as well as a range of beverages (including coffee, tea, orange juice, fruit punches and lemonade). Transport ships readily restock the ISS, so fresh supplies of food are only a moment away.

On their journey to Mars, astronauts on the *Orion* spacecraft will also need a robust diet to keep healthy and alert, but the spacecraft has limited room to accommodate the food and supplies necessary for the mission, and there will be no resupply ships to help them along the way. All the food they need will have to be stored onboard, and a crew of four on a three-year mission to Mars eating only three meals each day would need to carry more than 11,000 kilograms (24,250 lb) of food.[26] Adding to the challenge is how to prepare food for a three-year journey that will remain fresh while retaining its nutritional value and keeping bulk down (more weight requires more fuel and greater expense).

One development to cut down on bulk is being devised at NASA's Human Research Program, a variety of low-mass but calorically dense breakfast bars (like the energy bars we eat on Earth, but ones that will keep).[27] The goal is to have a number of food bars to select from in a variety of flavours, such as orange cranberry or barbecue nut, for their first meal of the day, reducing the amount of space and storage the breakfasts require. For lunch and dinner *Orion* astronauts may eat items from multiple thermostabilized (heat processed to destroy bacteria and other organisms so they can be stored at ambient temperature) or rehydratable packages as eaten by the ISS astronauts.

Crew members inside HERA – NASA's three-storey habitat at Johnson Space Center designed to serve as an analogue for the isolation and remote conditions in exploration scenarios – have already tested the food bars and provided helpful feedback on taste and texture, which will ultimately help prevent food fatigue and aversion among astronauts travelling to Mars. Scientists are also looking at packaging food items to keep them edible and nutritious in conditions where there are temperature fluctuations, such as the surface of Mars.

Turning the Red Planet Green

While the early Mars colonists will initially rely on prepackaged food for survival, they will also have stowed with them seeds to grow their own food. Gardening on Mars not only will provide colonists with extra nutrition, but can increase morale by sprouting a little bit of Earth on Mars. Already the Vegetable Production System (Veggie), a deployable plant growth unit enabling space gardening and space plant biology experiments on the ISS, produces salad-type crops that provide the crew with a palatable, nutritious and safe source of fresh food, while offering relaxation and recreation.

Once camps are established on Mars, future pioneers are more likely to produce most of what they eat from local resources. But Martian regolith, which recent research has shown contains some toxic chemicals (perchlorates) in the harsh UV surface environment that can kill certain types of bacteria and plants, is another challenge to overcome.[28] Despite this, as long as the toxins are flushed out, nutrients are added and the soil is worked, there is a chance that we can grow vegetables on Mars. A nearly month-long, NASA-sponsored pilot study at Florida Tech by Drew Palmer, a professor of biochemistry and chemical ecology, and Brooke Wheeler, an ecologist and professor in the College of Aeronautics, grew lettuce plants in three conditions: one in simulated Martian regolith, one in a simulated Martian regolith with added nutrients, and one in potting soil. These trays of lettuce were grown in a chamber with a controlled lighting and temperature setting. The lettuce grown in the Mars-like soil simulant tasted the same as the others. The only difference was that it was marked by weaker roots and a slower germination rate compared with the stronger roots displayed in potting soil plants.

'Martian gardens' have since cropped up elsewhere and experiments are being conducted on a variety of produce that could thrive on Mars, including lettuce, spinach, carrots, Chinese cabbage, Swiss chard, tomatoes, green onions, radishes, bell peppers, snow peas, strawberries and fresh herbs. In a separate experiment, scientists at Kennedy Advanced Life Support Research are working on the Prototype Lunar/Mars Greenhouse Project, to develop a closed-loop hydroponic system that uses plants to scrub carbon dioxide, while providing food and oxygen. As crops are grown, the system recycles water, recycles waste and revitalizes the air.

NASA researchers are also interested in the role cyanobacteria (commonly known as blue-green algae) can play in food generation and more on Mars. These microorganisms, which represent the earliest-known form of life on the Earth, are capable of producing oxygen as a by-product of photosynthesis, using sunlight to make

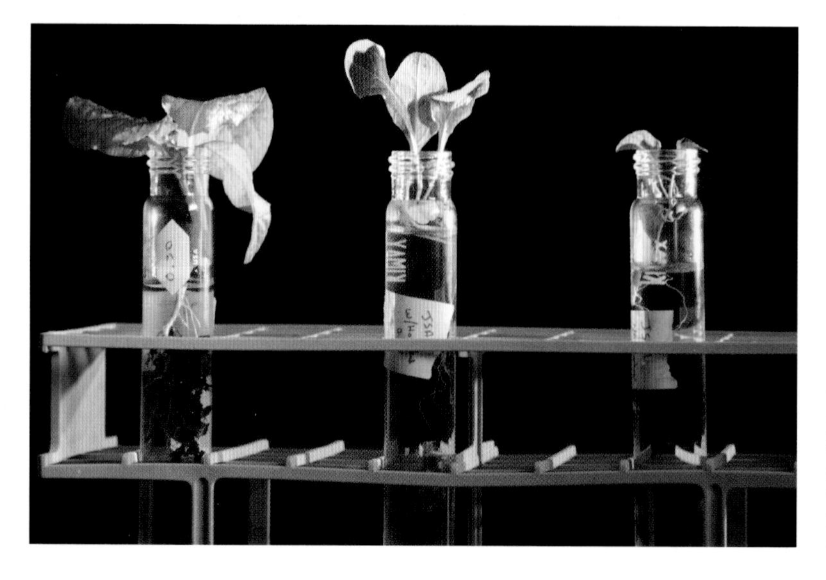

Plants were grown in a preliminary experiment comparing (left to right) potting soil, regolith simulant with added nutrients, and simulant without nutrients.

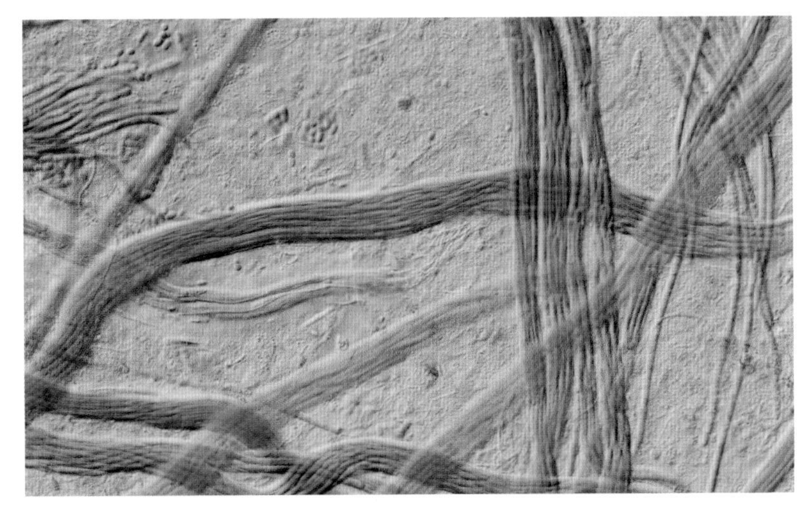

Light microscope view of blue green trichomes (in bundles) of the cyanobacterium *Microcoleus chthonoplastes*, collected from the Exportadora de Sal saltern system in Guerrero Negro, Baja California Sur, Mexico.

food and energy. While most terrestrial plants and microorganisms are unable to grow in the barren Martian regolith, a strongly UV-C radiation-resistant cyanobacterium has been discovered on Earth in conditions that most resemble Mars.

Some species of cyanobacteria are edible, can be grown on a large scale and could provide a nearly complete nutritional and high-protein source.[29] One species, *Arthrospira platensis*, is already used on Earth as a food supplement. When optimally produced it can produce twenty times more protein per hectare than soya on Earth. Cyanobacteria may also be biogenetically engineered to produce sugars. Cyanobacteria are already used on Earth as a main food source for the larvae of many species of fish, crustaceans and shellfish, which could lead to a dense aquaculture of these species in a controlled habitat on Mars.

Waste Not

Once on a journey to Mars, astronauts will throw nothing away, including human waste. How to turn such waste into food, oxygen and water is the subject of the ESA's Micro Ecological Life Support

System (MELiSSA) project being conducted at a plant outside Barcelona, Spain.[30] MELiSSA researchers are using communities of cyanobacteria to break down and recycle human waste (and the inedible parts of plants) so that it can be used to fertilize the soil in which to grow plants. The MELiSSA recycling system will consist of five separate, but interconnected, compartments, in three of which waste will be progressively broken down by different fermentation processes.[31] In the fourth compartment, algae or plants will grow to produce food, oxygen and water for the consumers in the fifth compartment.

At Wageningen University & Research in the Netherlands, experiments are being conducted on rucola (rocket) grown in Mars-simulated soil, provided by NASA, to which earthworms and pig slurry have been added.[32] The researchers discovered that not only does the manure stimulate growth in the simulated Martian regolith, but earthworms can reproduce in the soil created from it. Future humans on Mars will need a sustainable closed agricultural ecosystem, and earthworms may play a crucial role in this system as they break down and recycle dead organic matter. On Mars the pig slurry will be replaced with human faeces and urine to fertilize the soil.

A team of NASA scientists from Ames and Glenn Research Centers are also working on a device that could be used to incinerate any waste on long-duration missions.[33] Called the Vortical Oxidative Reactor Technology Experiment (VORTEX), the device uses variable-direction air currents to suspend waste particles indefinitely and incinerate them. Its ability to function in variable-gravity environments means that it could also be used to recycle waste into fertilizer on Mars.

Keeping the Dream Alive

The remarkable similarities between the Earth and Mars that earlier observers thought existed have proved to be illusory: Mars is not another Earth. We have long since awakened from the dream of Mars as an abode of intelligent life and have now accepted it for what it really is – a world of hidden wonders that has the potential to harbour microbial life, even if this eventually needs a little help from us. Certainly, for all its storied history, the study of Mars has really only just begun, and the future promises to be even more exciting than the past.

Establishing a human colony on Mars is now more than a dream. True, if we focus our attention only on the surface of facts that portray Mars as an inhospitable destination, we can easily suppress our desires to colonize the planet. But thanks to the collective vision of space agencies – in partnership with an expanding sphere of highly motivated entrepreneurs with foresight, flourishing finances and heaps of imagination – we already are well on the way to forging humanity's first intimate bond with the Red Planet. Once a flight of imagination, a journey to Mars is today within our grasp.

Elon Musk, the South African-born American technology entrepreneur, billionaire and founder of Tesla Motors and Space Exploration Technologies (SpaceX), is also keeping the dream alive in the private sector, believing that SpaceX will overtake NASA by being the first to sell space tourist seats on missions to Mars at a price of perhaps a couple of hundred thousand U.S. dollars per space traveller. He also estimates that humans could establish a city on Mars as early as 2050.

Musk has already set the foundation. SpaceX's *Falcon 1* became the first privately developed liquid-fuel launch vehicle to orbit the Earth, and NASA awarded it with contracts to carry cargo and crew to the ISS aboard the free-flying *Dragon* spacecraft: the first unpiloted *Dragon* flight launched successfully with the ISS on 3 March 2019. In 2018 SpaceX beat the world record for the most commercial rocket launches in a year, achieving twenty successful launches. It also developed other impressive aerospace systems: *Grasshopper*, a small self-landing test rocket; *Falcon 9*, a workhorse reusable orbital-class launcher that could have a lifespan of three hundred launches over five years; and *Falcon Heavy*, a super-heavy-lift launcher (and the world's most powerful launcher).

In February 2018 the maiden flight of the *Falcon Heavy* rocket boosted a roughly 2-tonne Tesla Roadster (with a mannequin driver named Starman decked out in a spacesuit) towards Mars at 40,000 km/h (25,000 mph). SpaceX said that launching a satellite or other valuable spacecraft was not an option, given the risks inherent in maiden flights. The red electric Roadster sailed past the Red Planet in November 2018 in its orbit around the Sun; present calculations show that it may be destined for a fiery re-entry into the atmosphere of either Venus or Earth within the next million years. The flight was, in a sense, a first step towards fulfilling one of its mantras: 'making life on Mars a reality in our lifetime'.

After the Roadster swung past Mars, Musk told the American financial website *Business Insider* that he is on track to launch people to Mars by 2024, although he stresses that no dates are firm.[34] For the task, Musk's current thinking (and this may change) is that he will use the Big Falcon Rocket, a 118-metre-tall (387 ft) two-stage system consisting of a Big Falcon Booster (called Super Heavy), roughly 70 metres (230 ft) tall and powered by 35 raptor engines, and a 55-metre-long (180 ft) second stage Spaceship (called *Starship*) powered by six main Raptor engines, a new SpaceX design that can provide a combined 5,400 tonnes of thrust. In January 2019, SpaceX completed assembly of the test-flight version of this Mars-colonizing rocket. The Starship prototype made its initial flight test, a 'hop' of around 20 metres (66 ft) altitude, in late July 2019. It made its second 'hop' the following August, reaching an altitude of around 150 metres (490 ft) and landing around 100 metres (330 ft) from the launchpad.

Business Insider outlines the scenario as follows. Each ship would first fly into orbit around Earth, a flight that would use up most of its fuel. Then several other tanker spaceships would launch to fill the vehicle with enough fuel to reach Mars, perhaps in late 2022

The last photo of 'Starman' in Elon Musk's red Tesla Roadster as it flies towards Mars orbit. Earth is the bright crescent seen in the background.

or early 2023. These first uncrewed missions will be full of cargo and machines needed for humans to build facilities that can generate power, gather water, bottle up the thin Martian air, and turn those raw resources into methane fuel and oxygen for return launches back to Earth. They will also confirm water sources, determine landing hazards for future missions and set up the initial infrastructure for the coming crewed missions. SpaceX plans to launch its first passengers on a journey around the Moon in 2023 to prove that the Super Heavy system works, before launching customers to Mars in 2024 with 100 tonnes of supplies. A year later it is possible that bootprints will appear in Martian soil.

The hardships explorers will have to face are real (some would even say formidable); but so too was once the summit of Everest or placing humanity's first foot on the Moon. But rather than 'summiting' Mars by imprinting a foot on it, or more than exploring the planet, which rovers have already been doing for decades, we want to farm it, to set up shop, if you will, to colonize the planet and create a new legacy for humanity. While some would argue that we are still decades away from achieving this, every year brings us closer to the next summit in the spirit of adventure: Mars. It is

Cover of the July 1948 issue of *Astounding Science Fiction*, showing vertical landing rocket.

A SpaceX visualization of its spaceship landing on Mars. The spaceship is designed to do powered landings, meaning it uses its engines to lower down to the surface – so reminiscent of the rocketships depicted in early science-fiction movies.

difficult to assess what technology has in store for us in the decades to come, or what humans themselves will be ready to endure once they commit to the task.

The Ethics of Colonizing Mars

Yearning for the human exploration of Mars is good, but we need to balance that desire with ethical considerations, as the author Michael Meltzer has warned: 'Careless planetary exploration in the present could forever obfuscate the answer to a vital question: Are we Earthlings alone in the universe?'[35] As future astronauts sail forth to Mars, the potential exists to contaminate relevant environments and possibly affect indigenous life adversely and irreversibly.

In 2012, for instance, researchers catalogued 298 strains of extreme bacteria that were able to survive the sterilization process in ESA clean rooms.[36] Although it is not suspected that robotic visitors have yet actively contaminated anywhere on Mars, terrestrial microbes may lie dormant in the soil. The possibility of contamination increases greatly if astronauts set foot on the Red Planet, especially with all their needs and support systems.

To address these concerns, NASA sponsors or co-sponsors interorganizational meetings and studies to map out how planetary protection requirements should be implemented during human missions, and what contamination control standards should apply to human explorers. Progress continues to be made, but many questions remain unanswered. What level of protection should we take in preserving a planetary environment? Should we view it as we do our terrestrial environment? The United Nations Educational, Scientific and Cultural Organization (UNESCO) and space science researchers have posited several questions regarding the issue, which depends on whether we view it as we do our terrestrial environment. Do we have an ethical obligation to preserve a planetary environment to the same degree that we seek to protect our Earth's environment? Does this obligation hold, even if there is no life on a planet? Since environmental ethics seek to benefit and enhance life, do we have an obligation to see that terrestrial life expands onto lifeless planets? Meltzer has pondered this quandary:

> The ancient Greek philosophers interpreted the study of ethics to be a search for answers to the question, how should we live? Such a question has value today, even as it did millennia ago, because as human beings, we must sometimes make decisions based on our own consciences and moral judgments in addition to simply obeying the law.

The question now, he says is, 'how shall we, the people of Earth, act in space? And in particular, how shall we act on bodies that may contain life?'[37]

EIGHT

The Lilliputian Moons of Mars

In 1877 a favourable opposition of Mars attracted the attention of U.S. Naval Observatory astronomer Asaph Hall (1829–1907), who decided to make a 'careful search' for any moons with the observatory's 66-centimetre (26 in.) refractor.[1] Starting in early August, a couple of faint objects some distance from the planet immediately captured his attention, but these turned out to be stars (moons of Mars would move against the fixed stars at the same rate as the planet). By 10 August Hall had begun looking closer to the planet by keeping the planet just outside the field of view and 'turning the eye-piece so as to pass completely around the planet'. On the night of 11 August he noticed a dim object near the planet, but barely had time to secure an observation of its position, when, as he described in a letter dated 28 December 1877, 'fog from the Potomac River stopped the work. Cloudy weather intervened for several days.' Hall succeeded again on 16 August when his observations confirmed that this suspect was indeed moving with the planet against the stars.

The next night, as Hall was 'waiting and watching' for his newly discovered satellite of Mars, he discovered another one nearer to the planet. For several days that inner moon perplexed him:

> It would appear on different sides of the planet in the same night, and at first I thought there were two or three inner

moons, since it seemed very improbable to me, at that time, that a satellite should revolve around its primary in less time than that in which the primary rotates. To settle this point I watched this moon throughout the nights of August 20 and 21, and saw that there was, in fact, but one inner moon, which made its revolution around the primary in less than one-third the time of the primary's rotation, a case unique in our solar system.[2]

Asaph Hall (1829–1907).

Of the various names proposed for the moons of Mars, Hall liked best those suggested by the English chemist Henry George Madan, who taught science at Eton College, near Windsor, England: Phobos (flight) for the inner moon, and Deimos (fear) for the outer moon. These names, according to Madan, represent the sons of Ares (Mars) who accompanied their father into battle, driving his chariot and spreading fear in his wake, as described in William Cullen Bryant's translation of Homer's *Iliad*:

> [Ares] spake, and summoned Fear and Flight to yoke his steeds, and put his glorious armor on.

Hall's discoveries are a testament more to the observer than to the telescope, as historical giants had searched in vain before him, some with larger telescopes: William Herschel searched unsuccessfully for Martian moons in 1783, which was an opposition year almost as favourable as 1877; Lord Rosse had failed to detect the moons with his 1.8-metre (5.9 ft) Leviathan reflector in 1845 (during another

(Chloe) Angeline Stickney Hall (1830–1892).

favourable opposition); William Lassell missed them with his 61-centimetre (24 in.) and 1.2-metre (4 ft) telescopes; and the moons of Mars also eluded Heinrich Louis d'Arrest, who scrutinized the planet's vicinity in 1862 with the 25.4-centimetre (10 in.) refractor at Copenhagen Observatory.

Hall also had to thank his wife Angeline, a suffragette and mathematician.[3] On the night before he discovered Deimos, the observing conditions were so dismal that Asaph anchored the telescope, closed up the dome and went home, where he complained that further searches would be useless.[4] Angeline, however, urged him on, insisting he return to the telescope the next night and discover the moons. In return he dedicated these discoveries to her, as described in the poem that forms the epilogue of the biography by their son Angelo:

> The bride of Science she; and he the groom
> She wed; and they together loved and learned.
> And like Orion, hunting down the stars.
> He found and gave to her the moons of Mars.[5]

Beyond Angeline's encouragement, Asaph's imagination and willingness to doubt conventional wisdom ultimately led him to fame. 'All that was needed,' Hall wrote, 'was the right way of looking, and that was to get rid of the dazzling light of the planet'[6] – especially as he was using the world's largest refracting telescope at the time. As often happens, however, once astronomers learned of the discovery and knew where to look, they began seeing the moons in smaller and smaller telescopes. Despite their minuteness, Phobos and Deimos are surprisingly bright whenever Mars reaches opposition. On better-than-average oppositions, these satellites can shine as brightly as magnitudes 11 and 12.0, respectively, for a few days around closest approach to Earth. If Mars could somehow be removed from the sky, a skilled observer at a dark site could easily pick out these moons with a pair of 11 × 80 binoculars mounted on a tripod.

After the announcement of the discovery, good observations of both satellites were made at the Harvard College Observatory with the 38-centimetre (15 in.) refractor by Edward C. Pickering and his assistants, while Hall and his colleagues John Eastman and Henry M. Paul glimpsed Deimos with the Naval Observatory's 24-centimetre (9.5 in.) refractor. In short order they were seen with ever smaller instruments, including Wentworth Erck's detection of Deimos with a 19-centimetre (7.5 in.) refractor on 3 September 1877.[7]

Lilliputian Worlds

As electrifying as Asaph Hall's discovery was to the general public of 1877, the story had a familiar ring. Almost exactly a century and a half earlier, the Anglo-Irish satirist Jonathan Swift had described the advanced state of scientific knowledge on the fanciful flying island of Laputa. Swift came remarkably close to describing the true state of the Martian moon system:

> They have likewise discovered two lesser stars, or satellites, which revolve about Mars; whereof the innermost is distant from the centre of the primary planet exactly three of his diameters, and the outermost, five; the former revolves in the space of ten hours, and the latter in twenty-one and a half; so that the squares of their periodical times are very near in the same proportion with the cubes of their distance from the centre of Mars; which evidently shows them to be governed by the same law of gravitation that influences the other heavenly bodies.[8]

The passage reveals Swift's knowledge and understanding of Kepler's Third Law of planetary motion: 'The square of the orbital period of a planet is proportional to the cube of the semi-major axis of its orbit.' He also knew how to make use of it: Hall found the moons' distances to be 1.4 and 3.5 diameters from Mars's centre, compared to Swift's 3 and 5; their periods are 7h 39m and 30h 18m, as against Swift's 10h and 21½h.

Ironically, Kepler's own beliefs may have planted the seed in Swift's mind that Mars has two moons. The roots of this belief have been traced to a letter Kepler wrote to Galileo, in which he said, 'I am so far from disbelieving in the discovery of the four circumjovial planets, that I long for a telescope, to anticipate you, if possible, in discovering two round Mars (as the proportion seems to require)'. As Owen Gingerich explains, this belief led Kepler 'into a curious trap'.[9]

In 1610 Galileo observed globe-like appendages on either side of the planet Saturn and believed them to be other planets (they were the rings, unresolved). To keep his finding secure, he announced the discovery to the world as a Latin anagram, whose solution, *Altissimum planetam tergeminum observavi*, can be translated 'I have observed the most distant planet [Saturn] to have a triple form.' Kepler, however, misconstrued the anagram and offered a wrong solution, *Salue umbistineum geminatum Martia proles*, meaning 'Hail, twin companionship, children of Mars.'[10]

While Kepler's surmise may have inspired the remarkable passage in *Gulliver's Travels*, as the English researcher N. Rosewier stresses, Newton's 'Origins' asserts that 'all things being equal, the smallest planets have much higher density'. If one assumes that Mars' density is 22 times that of Jupiter (since Jupiter's diameter is about that many times the diameter of Mars) and applies Kepler's third law, then one can obtain Swift's result.[11] Jonathan Swift, though clever, was neither clairvoyant nor original in his prediction that Mars had two moons. It would have been an interesting coincidence also, getting the answer right for the wrong reasons, but nothing extraordinary.

Nevertheless, by solar system standards the moons of Mars are without question Lilliputian worlds. Phobos, the larger of the two, measures only 22 kilometres (14 mi.) across (only half the width of Greater London); the moon is so small that a 68-kilogram (150 lb) person standing on its surface would weigh only 57 grams (2 oz). Deimos is a little more than half that size (13 kilometres (8 mi.)). If we could transport Phobos and Deimos to our own Moon, they would fit comfortably inside the crater Copernicus, with room enough for two more moons of similar size.

Not only are Phobos and Deimos among the solar system's tiniest moons but no other moons are known to orbit their primary so closely. Rising in the west and setting in the east, Phobos whips around Mars three times a day from an orbit only 6,000 kilometres (3,730 mi.) above the Martian surface. Tiny Deimos takes thirty hours to complete an orbit from its height of 20,000 kilometres (12,500 mi.). If these moons circled Earth at the same distance as they orbit Mars, Phobos would pass below, and Deimos by, the average positions of our GPS satellites (~20,000 kilometres above the Earth's surface). Even though Mars is smaller than the Earth, the orbital periods of its moons are nevertheless remarkably short, and they would not be visible from all vantage points on Mars. An astronaut standing near Mars's pole would see neither moon, as they would orbit the planet below the horizon. To an astronaut standing near the equator, Phobos would remain visible for only about five-and-a-half hours (between rising and setting), while Deimos would sail across the sky much more slowly, taking about sixty hours to rise and set.

For nearly two decades after Hall's discovery the Martian moons were scrutinized visually, with a primary focus on making precise positional measurements. A great number of post-discovery observations are attributable to Russian astronomers at the

The moons of Mars compared to the crater Copernicus (the largest crater visible at left) on Earth's Moon.

Pulkova Observatory. In December 1896, for instance, S. Kostinsky used the 33-centimetre (13 in.) astrograph there to take the first photographs of Deimos. As that opposition and the years following were unfavourable for photographing 'these small celestial bodies', Kostinsky tried again in 1909 and succeeded again, this time capturing both moons:

> Already on 30 August 1909 I found traces of the satellite Phobos at its eastern elongation, on two tests of Mars made with exposures of 10 and 20 minutes duration. On the nights of 2, 9, 13, 14, 16 and 21 September I obtained a series of images

> of Phobos at both elongations, some of which are quite measurable. On the pictures of 13 and 16 September we can also see very good images of Deimos . . . As far as I know, our photographs of the Mars satellites are the first.[12]

Karl Hermann Struve (1854–1920), the son of Otto Struve, and grandson of Friedrich Georg Wilhelm von Struve, founder of the Pulkovo Observatory, developed the first theory of the motion of Martian satellites in 1911. Based on his extremely accurate observations from 1877 to 1909, Struve calculated the elements of the moons' orbits and successfully predicted their positions relative to Mars as seen from Earth.[13] The ephemerides in *The Astronomical Ephemeris* were based on his results, though they were later refined to account for tidal acceleration (more on this later).

The first visual photometric attempts at studying the two moons were performed between 1887 and 1892 under the direction of Edward C. Pickering at Harvard College Observatory.[14] Pickering's technique was laborious. As Mars was several hundred thousand times brighter than the satellites, direct comparisons were not possible. Instead, the astronomer first had to visually compare the brightness of the satellites directly to the light from Mars coming through a pinhole in a piece of foil placed over the planet's focal point. From the measured diameter of the pinhole and the photometer reading, the difference in magnitude between the satellite and Mars was derived. From this magnitude difference and the absolute magnitude of Mars, the absolute magnitude of the satellite was obtained. When Pickering combined this magnitude with the percentage of sunlight reflected from the surface of Mars (albedo) he found diameters of 12 and 10 kilometres (7.5 and 6 mi.) for Phobos and Deimos, respectively, which were remarkable results for such an onerous procedure.[15]

The Dutch-American astronomer Gerard Kuiper (1905–1973) performed the first photoelectric photometry of the satellites in the 1950s. His work led not only to more accurate size estimates for the satellites (21 kilometres (13 mi.) for Phobos and 13 kilometres (8 mi.) for Deimos), but the first albedo measurements (0.06–0.07). These extremely low values provided us with the first hint that the surfaces of Phobos and Deimos resemble carbonaceous chondrites, a rare type of primitive stony meteorite that contains a significant proportion of dark, carbon-rich materials. As these materials are commonly found in some asteroids, astronomers began to suspect that Mars had captured its two moons early in its history.

The Spacecraft Era

It was not until spacecraft began visiting Mars that the true nature of the moons materialized. Shortly after *Apollo 11* landed on the Moon (24 July 1969) NASA's *Mariner 7* arrived at Mars and snapped the first image of Phobos, which appeared as a minute silhouette against the Martian surface. From this data, American astronomer Bradford Smith found Phobos to be an oval object larger than previously thought (22.2 km (13.8 mi.) wide) and confirmed that it reflected only about 6.5 per cent of the sunlight striking it (lower than that known for any other body in the solar system).[16] The data further advanced the current thinking at the time that Phobos did not form *in situ* by accretion around primordial Mars but was a captured object.

NASA's follow-up *Mariner 9* mission, the first artificial satellite to orbit another planet, captured the first images of Phobos showing surface details. The images, taken between 1 December 1971 and 2 February 1972, unmask Phobos as a heavily cratered potato-shaped world with two enormous impact structures, dubbed Stickney (after Angeline Stickney Hall) and Hall (after Asaph Hall): Stickney takes a large bite off one end of the moon, while Hall overlaps the position of the moon's South Pole.

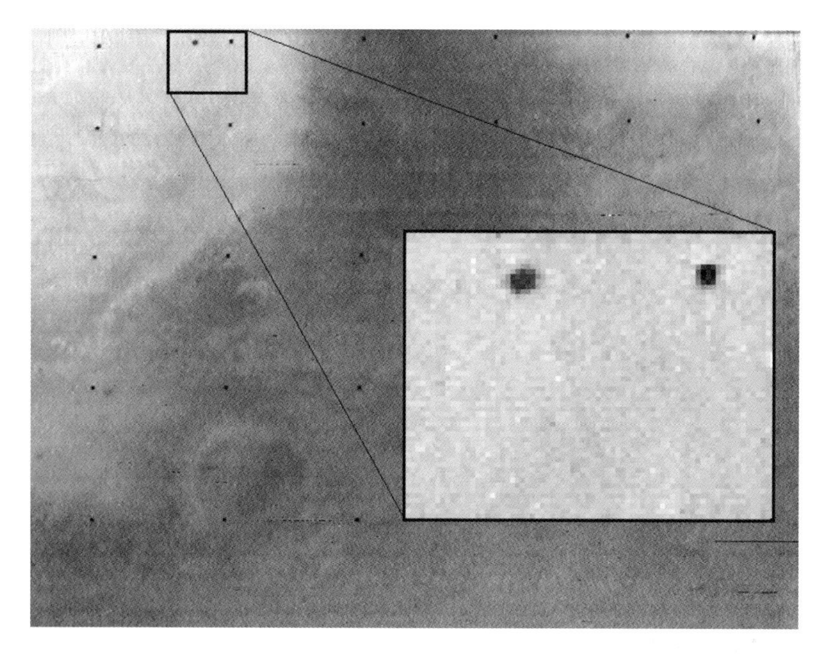

This *Mariner 7* image is one of the first spacecraft photos of Phobos.

Mariner 9 was also the first spacecraft to capture detailed views of Deimos. Its images portrayed this smaller irregular world as having a strangely smooth terrain with a long linear ridge and two enormous craters, named in honour of Swift and Voltaire. (In his short story *Micromégas* (1752), which tells of a voyage by an inhabitant of a planet orbiting Sirius to the planet Saturn, Voltaire also writes about Mars having two moons: 'Upon leaving Jupiter they traversed a space of around one hundred million leagues and approached the planet Mars, which, as we know, is five times smaller than our own; they swung by two moons that cater to this planet but have escaped the notice of our astronomers.'[17]) Suddenly the two specks of light discovered by Hall nearly a century before had blossomed into new worlds with their own geography; indeed, cartographers at the U.S. Geological Survey used the *Mariner 9* images to create the first two-dimensional maps of the Martian moons (though portions of them had yet to be seen).

In the summer and autumn of 1976 NASA's *Viking* 1 and 2 orbiters flew within a few hundred kilometres of the moons' surfaces – then as close as 100–300 kilometres (60–185 mi.) in February and May 1977 – revealing some unexpected surface features. Phobos, for instance, is dominated by sharp, fresh-looking craters of all sizes and a vast network of linear grooves, some up to tens of kilometres long and hundreds of metres across. These images suggested that the grooves, filled with craters or pits (or both), all lead to Stickney. If the grooves formed at the time of the Stickney impact, researchers wondered, they may be surface manifestations of deep fractures hidden beneath the dust. The *Viking* craft imaged no such features on Deimos.

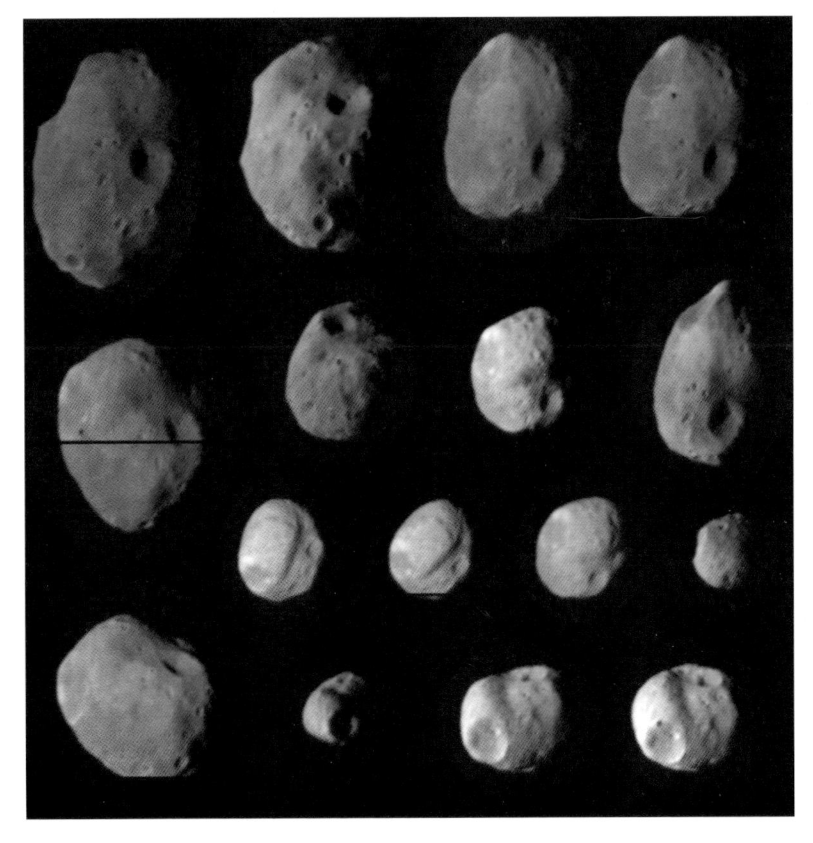

Sixteen views of Phobos from *Mariner* 9. The largest crater visible in many of these photos is Hall, which overlaps Phobos's South Pole. Stickney is most prominent in the last two images.

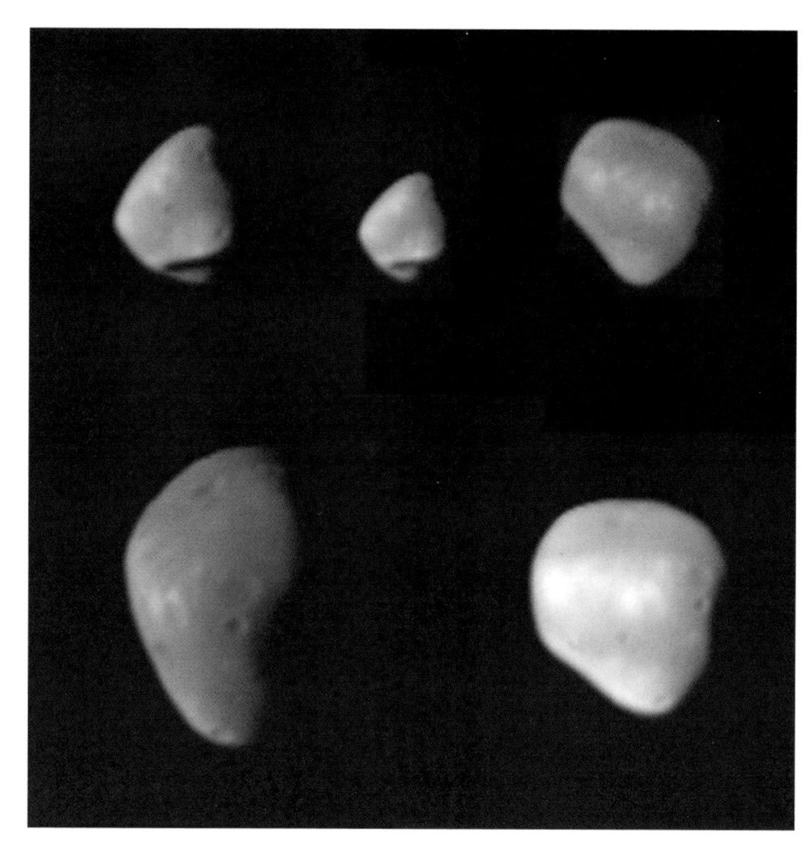

Five views of Deimos from *Mariner 9* between 20 December 1971 and 20 February 1972.

Viking data also explained the moons' low albedos: layers of low-reflectivity dust and broken rock (together known as regolith) cover the solid surface to depths of a few hundred metres for Phobos and at least 5 to 10 metres (16 to 33 ft) for Deimos. Cratering of the surface of Phobos continued during and after the formation of the regolith. On Deimos, however, it appears that the regolith continued to develop after the cratering subsided (so any craters measuring about 100 metres (330 ft) or less are partially filled or covered), which gives Deimos a much smoother appearance than Phobos when viewed at ranges of more than a few hundred kilometres.[18] Crater densities on both satellites are comparable to densities on the lunar highlands, a fact that suggests ages of up to a few billion years.

The Soviet Union initiated the first dedicated missions to Phobos in July 1988, launching *Phobos 1*, which suffered a terminal failure en route to Mars, and *Phobos 2*, which arrived at Mars on 30 January 1989, but was lost on 27 March 1989 while manoeuvring in Martian orbit, just before the spacecraft was to move within 50 metres (165 ft) of the moon and deploy its two landers.[19] Before contact was lost, however, *Phobos 2* transmitted 37 TV images of the moon to a resolution of roughly 40 metres (130 ft) and included some previously unseen areas, expanding the moon's geography. *Phobos 2* also detected a faint but steady outgassing from the moon, the nature of which remains a mystery because of the unfortunate loss of the spacecraft, but it may be related to water.

NASA's *Mars Global Surveyor* spacecraft made the next closest approach to Phobos in 1998, journeying within 1,080 kilometres (670 mi.) of the moon. The spacecraft imaged several new features associated with the 9-kilometre-wide (5.5 mi.) Stickney crater, including individual boulders near the crater's rim, theorized to be blocks of rocks ejected from the impact that formed Stickney.

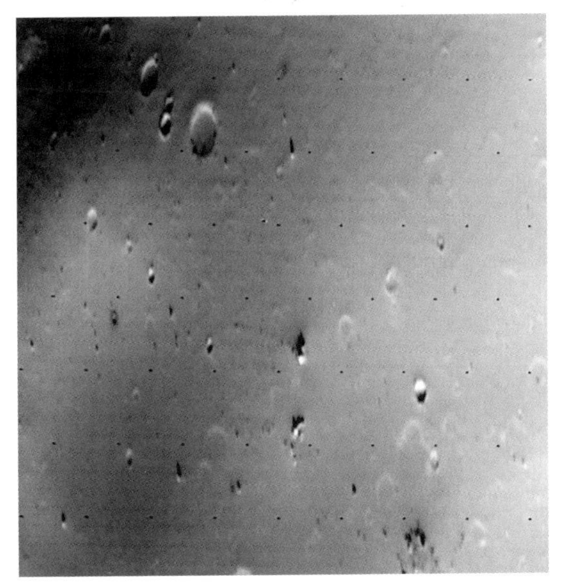

Left: *Viking 1* orbiter view of the southeastern quadrant of the anti-Mars side of Phobos. The South Pole is just off the lower right corner. The frame is about 17 km across. The craters on the limb are about 4 km across and hundreds of metres deep. Right: *Viking 2* orbiter image of Deimos. This was at the time one of the highest resolution images ever taken from an orbiting or flyby spacecraft. It shows the surface of Deimos from a distance of 30 km. The image covers an area 1.2 km by 1.5 km and features as small as 3 m across can be seen. Note many of the craters are covered over by a layer of regolith estimated to be about 50 m thick. Large blocks, 10 to 30 m across, are also visible.

Some of these boulders are enormous, more than 50 metres (165 ft) across. Aside from the previously observed grooves, the evidence of lighter and darker streaks on Stickney's far wall implies that the satellite is *heterogeneous* (made of a mixture of different types of materials). Data from the spacecraft's thermal spectrograph reveal that these streaks were most likely formed by meteorite impacts pounding the surface and causing landslides on the steep crater slopes.[20]

The craft also measured temperatures across the surface, showing intense differences depending on the terrain: surface temperatures hover around −4°C (25°F) in direct sunlight, before plunging to about −112°C (−170°F) in shadow. This difference is most likely caused by the powdery dust on the surface losing heat whenever parts of this airless world are thrust into shadow. The thermal spectrograph also provided a wealth of information about the moon's surface composition, including its surface regolith being composed of a silica/carbonate mixture, which is most likely being eroded by meteorite bombardment. It also found the presence

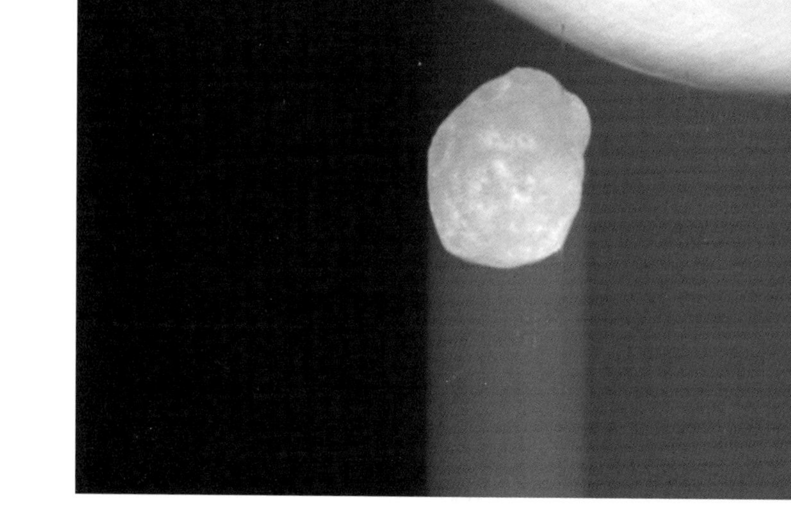

Raw TV image of Phobos seen with section of Mars, taken by the Soviet Union's *Phobos 2* spacecraft on 25 March 1989 from a range of 195 km – two days before contact with the craft was lost.

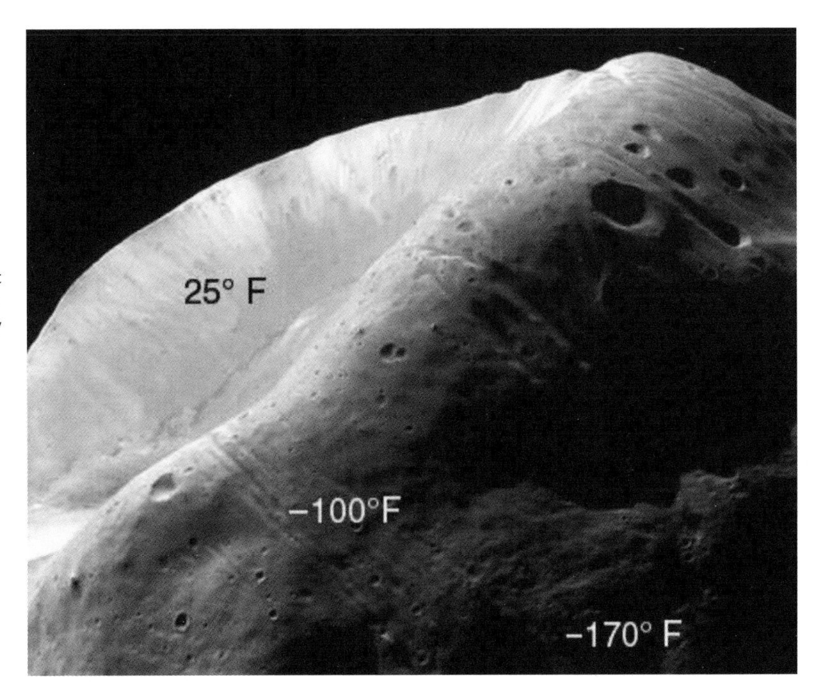

This *Mars Global Surveyor* image taken in August 1998 is one of the highest-resolution images ever obtained of Phobos (4 m per picture element). The giant crater Stickney is at the top; note grooves that scientists believed may be fractures caused by the formation of this crater that is nearly half the size of the satellite. Temperatures vary on Phobos, reaching highs of −4°C in direct sunlight, while shadow regions can be as cold as −112°C. (The temperatures here are shown in Fahrenheit.)

of water, which may be either a transient surface phenomenon or bound in minerals.[21]

The first impressive colour images of both Phobos and Deimos were transmitted by NASA's *Mars Reconnaissance Orbiter* on 23 March 2008, when the spacecraft approached to within 6,800 kilometres (4,225 mi.). Most striking is the rim of Stickney's crater, which appears bluer than the rest of Phobos. Based on analogy with materials on our own Moon, this could mean this surface is fresher and therefore younger than other parts of Phobos, or hasn't been exposed to space as long as the rest of Phobos's surface has.[22]

The colour views of Deimos show more subtle variations, with the smoothest areas appearing redder than those near fresh impact craters and over the moon's highest ridges. The colour differences are probably caused by exposure of surface material to the space environment, which leads to darkening and reddening. As with Phobos, the brighter and less-red surface materials on Deimos

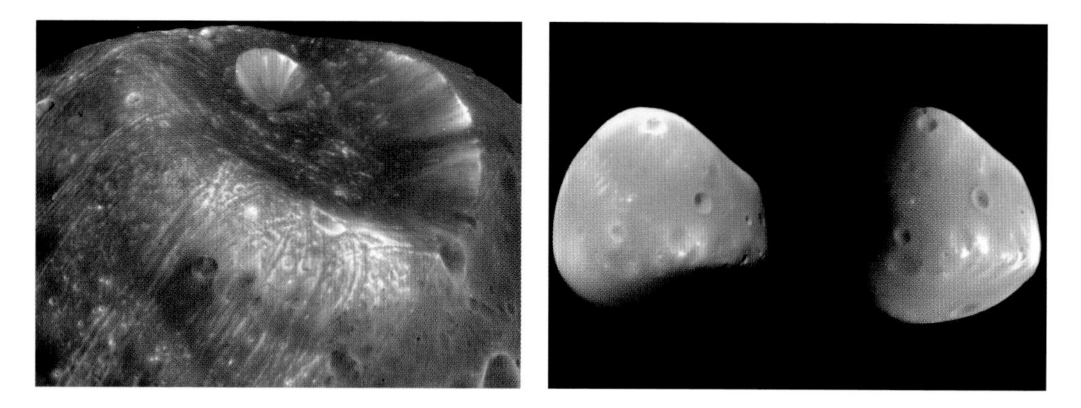

may have received less exposure to space due to recent impacts or downslope movement of regolith.[23]

Left: colour-enhanced *Mars Reconnaissance Orbiter* image showing how the material surrounding the giant crater Stickney appears blue-grey while the rest of the moon appears reddish. We can see features as small as 20 m in this image.

Right: Two colour-enhanced images of Deimos taken on 19 February 2009, showing details at 20 m per pixel.

Origins

While the jury is still out on the origins of the Martian moons, data from the European Space Agency's *Mars Express* mission are causing scientists to rethink much of what they previously surmised. One discovery came as a surprise. During several flybys of Phobos at progressively closer distances (the closest being only 97 kilometres (60 mi.)), *Mars Express* imaged the series of grooves and crater chains on Phobos, long thought to radiate from Stickney, finding that they are not related to that crater at all. Instead, they are centred on the highest point of Phobos's leading tip. It is possible that these grooves and streaks may have formed after violent impacts on the surface of Mars blasted material into space, some of which Phobos collided with as it orbited Mars. This scenario appears to be supported by the fact that all the resulting crater chains fade out towards the moon's trailing end. Nevertheless, the origin of the grooves is still being debated with two sparring theories that they were produced by ejecta thrown up into orbit from impacts into Mars, or that they are the result of the surface regolith slipping into internal fissures in Phobos.[24]

Another important revelation occurred in December 2013, after *Mars Express* performed the closest flyby to Phobos, skimming past the moon at a distance of only 67 kilometres (42 mi.). As it neared, the moon's gravity pulled the spacecraft slightly off course, allowing scientists to measure the moon's density, which they found to be unexpectedly low (only 1.86 grams per cubic centimetre). The results suggest that Phobos is most likely a porous body rather than a single chunk of solid rock, and may have large internal cavities accounting for about 25 to 35 per cent of the moon's total volume.

Spectral measurements of Phobos's surface composition further suggest the presence of sheet minerals, such as mica, talc and clay (particularly near Stickney). This implies that silicate materials interacted with liquid water in the past and adds the possibility that Phobos and Deimos are aggregates of material, not monoliths of rock and ice. These findings make it less likely that Phobos and Deimos are captured asteroids, although it does not rule out the possibility that they are captured achondrite-like objects (rare stony bodies consisting mainly of silicate minerals with the texture of igneous rock but with no mineral grains), which most likely formed on a planet with a distinct core and crust – like Mars.[25]

It is possible that Phobos and Deimos are second-generation solar system objects created by accretion after another body, with about one-third the mass of Mars, collided with the planet.[26] Such a fantastic impact would have ejected material into space creating a ring of debris around Mars, which subsequently coalesced into the two moons. If true, we do not know why this ring of debris formed into two bodies at such disparate distances from Mars.

RIP, *Phobos?*

Phobos (Fear) is appropriately named, as the moon is on an inward death spiral. Its orbit is decaying by about 2 metres (6.5 ft) per century, which means that intense tidal forces could rip it apart as early as 30 million years from now, and Terry Hurford at NASA's Goddard Space Flight Center believes that process has already begun.[27]

As early as 1945 Bevan Sharpless's studies of Phobos's orbital motion at the U.S. Naval Observatory revealed that the moon is in permanent acceleration,[28] leading him to opine that Phobos is spiralling closer to the planet, and that in about 15 million years the satellite would collide with Mars and cease to exist. Now Hurford and his colleagues believe that the first sign of this failure is the production of the linear grooves on Phobos. While the commonly accepted theory is that the grooves were created by material ejected from Mars during a violent impact with the planet, new modelling by Hurford and colleagues supports the view that the grooves are more like 'stretch marks' that occur when Phobos gets deformed by tidal forces.[29]

That idea that Phobos is tearing apart was first introduced after the *Viking* spacecraft sent images of Phobos to Earth. At the time, however, Phobos was thought to be more or less solid all the way through. When the tidal forces were calculated, the stresses were too weak to fracture a solid moon of that size. Now that researchers are looking at Phobos more as a 'rubble pile' (barely held together by a mildly cohesive layer of regolith) than a stone monolith, the theory has returned. Such a structure has very little strength, causing its interior to easily distort. Thus the researchers think that tidal forces acting on Phobos can produce more than enough stress to fracture the surface. When they modelled such a malleable body under gravitational duress, the stress fractures predicted by their model line up very well with the grooves seen in images of Phobos.

Mars Express image of Phobos taken on 28 July 2008 during one of its close flybys, showing the curious grooves that appear like stretch marks, possibly indicating the moon is starting to tear apart.

This explanation also fits with the observation that some grooves are younger than others, which would be the case if the process that creates them is ongoing.

The irony is that if Phobos was indeed born from a ring of debris ejected into Martian orbit during a violent impact on the surface of Mars, Phobos, as it approaches the Red Planet, will be torn into a ring again. One way or another, within 50 million years it will either crash into Mars or break up and form a temporary ring around the planet.[30]

Quick Facts

Phobos

Orbit from Mars	9,378 km
Diameter	22.5 km
Mass	1.06×10^{16} kg

Deimos

Orbit from Mars	23,455 km
Diameter	12.4 km
Mass	1.47×10^{15} kg

Spying the Moons of Mars

Despite their elusiveness, Mars's two moons, Phobos and Deimos, are surprisingly bright whenever Mars reaches opposition. The problem is they are so close to brilliant Mars that the surrounding glare tremendously overpowers them, just as it is difficult to see stars with the naked eye next to a full Moon, even though they would be quite noticeable in the absence of the Moon. Mars is 200,000 times brighter than Phobos and 600,000 times brighter than Deimos.

One trick to seeing them is to remove the planet's glare by manufacturing an occulting bar that extends completely across the field of view. A temporary one is easy to contrive from a piece of a wire or a narrow strip of aluminium foil. Most eyepieces have a field stop, a circular metal ring, at the focal plane. Experience shows that a deep-blue or violet Wratten gelatin filter works equally well. Trim the filter to a semicircular shape and mount it in the eyepiece's focal plane so it masks half the field of view. Since Mars will shine dimly through this mask, you can more easily judge the distance and direction from the planet where you should look for each moon. Keeping the planet behind the mask is also easier, especially if the telescope has an imperfect drive or no drive at all.[31]

In 1988 Alan MacRobert described to the readers of *Sky and Telescope* how a similar modification could be made to a 6-millimetre (0.2 in.) eyepiece:

> Cut a strip of foil about 1 mm wide and 1 or 2 cm long, smooth it with a finger, and place a bit of tape across one end. After unscrewing the hollow chromed barrel from the eyepiece, tape the foil strip to the outside of the field stop so it extends across the eyepiece aperture. Then while sighting through the eyepiece, carefully nudge the strip with a pencil point until at least part of it is in sharp focus near the center of the field. That is where to hide Mars.[32]

Scores of amateurs took MacRobert's advice during the very favourable 1988 opposition and succeeded with the aluminium foil occulting bar. Most observers using 28-centimetre (11 in.) and larger telescopes called Deimos 'easy' when it reached greatest elongation. In September 1956, the famous Mars observer Tsuneo Saheki spotted Phobos with a 20-centimetre (8 in.) Newtonian at the Osaka Planetarium in southern Japan. The smallest telescope to show Deimos was a 15-centimetre (6 in.) f/15 Jaegers refractor at the University of Cincinnati Observatory in Ohio. Basil Rowe, who made this sighting on one very transparent night in 1988, said the moon was at the threshold of detection. He confirmed the sighting with a nearby 40-centimetre (16 in.) telescope. Other favourable Mars oppositions have brought similar tales of success.

No matter which method you use, it is always best to try for the moons near an eastern or western elongation (maximum separation) from the planet. Because Phobos and Deimos lie roughly east or west of Mars at their greatest elongations, rotate the eyepiece so that your home-made occulting bar runs roughly north–south. Then try to spot the dim moons in the planet's background glow.

Clearly, success in seeing the Martian moons with an amateur telescope depends on a multitude of factors working in concert: the moons must be near or at greatest elongation; the atmosphere must be stable enough to show sharp images at high power; and the planet should be masked. The telescope has to be of sound quality and its optics free of dust, which scatters light, and the observer must patiently follow the object to be sure it moves with the planet and is not a background star.

NINE

OBSERVING MARS

Mars is the only planet that cannot be seen every year, being visible about every two years for a period of several months. The period of Mars visibility (known as an *apparition*) can be either favourable or unfavourable. Mars always appears best in the sky at opposition (when an astronomical body is opposite the Sun as seen from Earth and rises when the Sun sets; for instance, when the Moon is full, it is at opposition). Next to Mercury, the orbit of Mars is the most eccentric (0.0934), so the distance between Earth and Mars changes with each successive opposition. When Mars is closest to the Sun in its orbit, it is at *perihelion*; when farthest away, it is at *aphelion*. *Perhelic oppositions*, then, are the most favourable to observe, because Mars appears not only at its brightest to the unaided eyes but the largest through our telescopes.

But these are rare events, occurring at intervals of fifteen or seventeen years, when the distance between the Earth and Mars is at a minimum (about 56 million km (35 million mi.)), and we are experiencing either northern hemisphere summer or southern hemisphere winter. The last perihelic opposition occurred in the summer of 2018, when Earth and Mars were separated by 57.7 million km (35.8 million mi.) on 27 July and the Red Planet was 24.1″ arcseconds in apparent diameter, the angular measure of an object's dimensions against the celestial sphere: 1° = 1/360 of a circle; 1′= 1/60 of a degree; and 1″ = 1/60 of an arcminute.

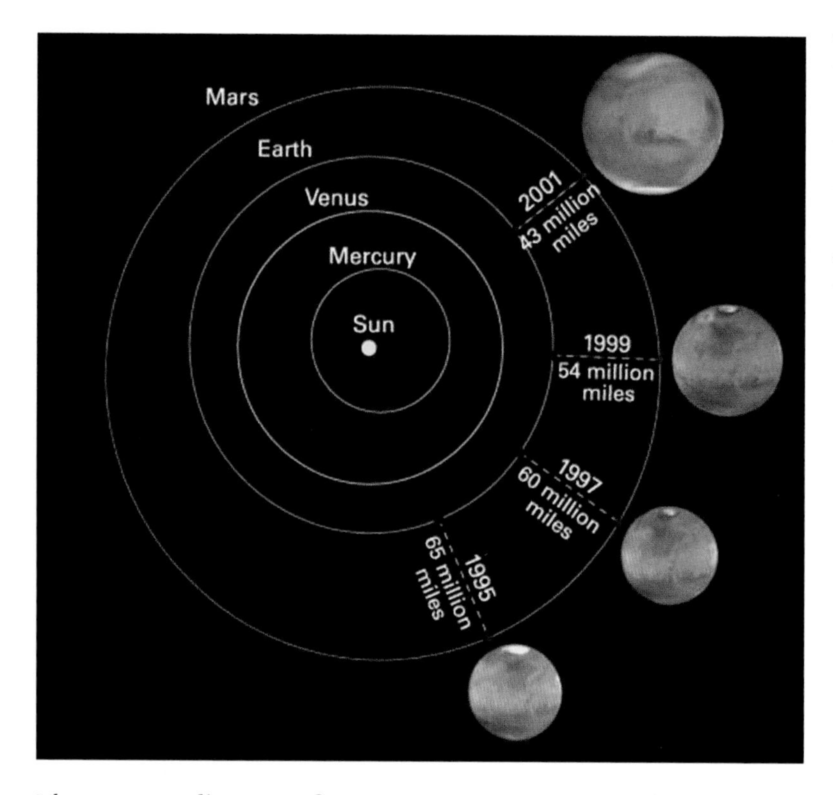

This illustration shows the relative positions of Earth and Mars at four oppositions. The images of Mars show the planet's apparent relative size at each opposition, as viewed by the Earth-orbiting Hubble Space Telescope. Orbits of the inner planets are to scale.

The apparent diameter of Mars at maximum is about 0.42′ or about 25″. The next time Mars will appear this large will be on 22 October 2035 (24.5″).

Unfavourable apparitions of Mars occur when Mars is farthest from the Sun in its elliptical orbit (*aphelion*), and the distance between the two planets at opposition is at a maximum (about 100 million km (62 million mi.)); aphelic oppositions always occur in northern hemisphere winter or southern hemisphere summer. The last aphelic opposition occurred on 3 March 2012, when Earth and Mars were separated by 100.8 million km (62.6 million mi.) and the Red Planet appeared 14″ in apparent diameter. During the best perihelic oppositions, Mars can shine at nearly magnitude –3 (almost as brightly as Venus); at its best aphelic oppositions, the planet shines at only magnitude –0.8 (about as bright as Saturn). Consequently,

Mars appears fifty times brighter when nearest Earth than when farthest away.

During its best oppositions, Mars can be sighted with the unaided eyes during daylight. The author's latest success came on 11 September 2018, when Mars was 48° above the horizon at 131° elongation from the Sun and 0.49 astronomical unit from Earth; and shining at magnitude –1.8. The next afternoon he sat in a chair and found Mars thirty minutes before sunset through binoculars, and sixteen minutes before with the unaided eyes. In Tomball, Texas, seventeen-year-old Lauren Herrington had success sighting Mars with the naked eye on 17 September 2018, as did Texas amateur Scott Harrington, who made his record naked-sighting of the Red Planet seventeen minutes before sunset on 19 September 2018.[1]

Although Mars is popularly known as the Red Planet, its colour can appear any variety of warm shade: from red to pink, orange to gold, or yellow. When an apparition starts, Mars is distant, appearing as a reddish second-magnitude star (about as bright as a star in the Plough or Big Dipper). As Mars approaches the Earth in its orbit, its brightness gradually swells to a maximum that rivals brilliant planet Venus. This change in brightness increases the perceived contrast between the planet and the night sky, resulting in a dramatic colour shift from blood red to golden yellow. The colour shift reverses once the planet moves away from Earth and its brightness wanes. Keeping records of the planet's naked-eye colour during a single apparition can lead to a better understanding of visual perception. It can also help us to better appreciate the influence Mars has had on humanity over the centuries.

Mars through the Looking Glass

Apparitions of Mars begin when the planet rises in the east shortly before the Sun; they end when the planet slips below the horizon shortly after sunset. During the several-month period of visibility, observers can use their telescopes to monitor the planet's phases, atmosphere and surface, carefully monitoring the world for any changes. But the task is challenging. During perihelic apparitions the planet appears seventy times smaller than does the full Moon. During Mars's aphelic apparitions, the Red Planet is twice smaller still; and when Mars makes its first or last appearance during an aphelic apparition, its disc can appear three hundred times smaller than the full Moon.

Most observers tend to monitor Mars around the time of opposition, when the planet is closest to Earth and largest in their telescopes; at the best perihelic oppositions, the apparent diameter of Mars reaches 25.1″. During best aphelic oppositions, the disc is much smaller, being only about 14″. Both oppositions have benefits and handicaps for Mars viewers. While the planet is larger during perihelic oppositions, it is also lower in the sky for northern hemisphere observers, where unstable air can cause imperfect views. The atmosphere of Mars is also very dusty at this time: the Martian dust storm season generally begins around late spring or

This composite image shows why observing Mars is a challenge, because even when largest it appears small through our telescopes, about the size of a moderately sized crater on the Moon.

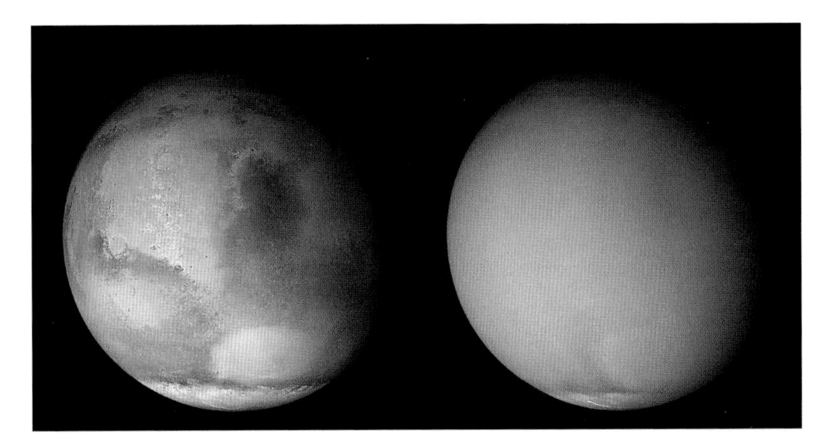

Two 2001 images from the Mars Orbiter Camera on NASA's *Mars Global Surveyor* orbiter show a dramatic change in the planet's appearance when haze raised by dust-storm activity in the south became globally distributed. The images were taken about a month apart. North is up.

early summer in the Martian southern hemisphere, when the planet is closest to the Sun. Airborne dust softens the view by lowering the contrast between the dark surface features and their surrounding desert regions. Occasionally, planet-wide dust storms can obscure the surface entirely from view.

During aphelic oppositions, Mars is not only much higher in the sky, but the Martian atmosphere is clearer, allowing unobstructed views of surface detail, which also appear in greater contrast. The downside is that the diminutive size of the planet requires a larger aperture to see fine detail (but not everyone is after that in their searches, so don't be dismayed). Also, while the Martian atmosphere is *nearly* dust free, cirrus clouds may be more prolific; in general, the contrast between the dark and bright surface markings are much stronger. Indeed, as Schiaparelli pointed out long ago, the size of the disc is less important than the transparency of the Martian atmosphere in determining the visibility of minor features.

If you want to make the most of your Martian studies, long-term studies throughout the apparition (even when the planet's disc is 6″ or 7″ of arc; the minimum size Mars can appear during an aphelic opposition is 3.5″) could reveal some time-dependent changes that may help astronomers answer some of the most interesting questions about Mars. Despite the prodigious number of spacecraft orbiting

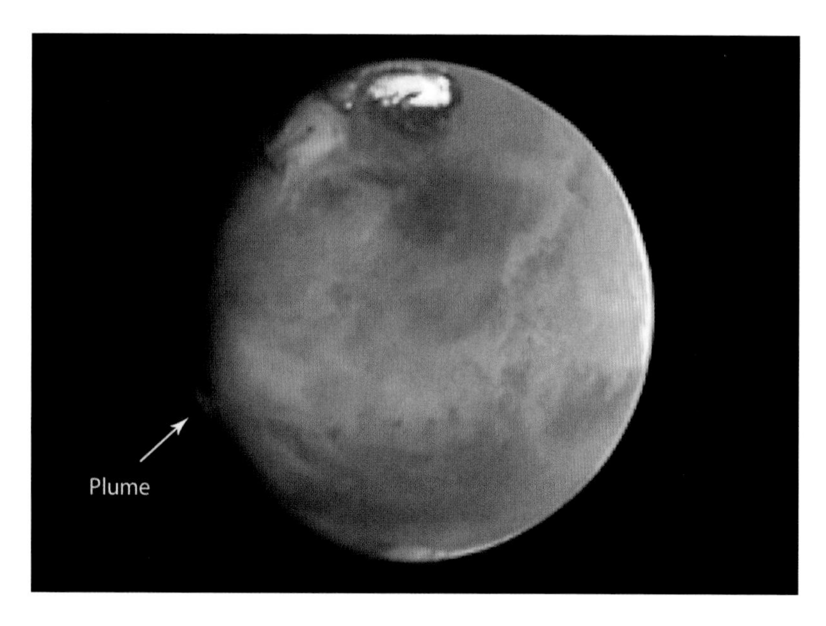

A curious plume-like feature was observed on Mars on 17 May 1997 by the Hubble Space Telescope. It is similar to the features detected by amateur astronomer Wayne Jaeschke in 2012, although it appeared in a different location.

Mars, amateur astronomers using backyard telescopes can still contribute to Mars science.

On the night of 19 March 2012, for instance, Wayne Jaeschke, a patent lawyer from West Chester, Pennsylvania, was taking images of Mars through his 20-centimetre (8 in.) Celestron telescope when he noticed a tiny cloud-like plume protruding nearly 200 kilometres (125 mi.) above the Martian surface. Alerting the astronomical community through message boards, he brought the feature to the attention of professional astronomers, who were baffled by the phenomenon. Jaeschke's message also prompted other amateur and professional Mars imagers to check their work, leading to eighteen additional images of the mysterious plume, as well as possible similar plumes found on archived images as far back as 1997, as recorded by the Hubble Space Telescope.

Two years after his discovery, Jaeschke co-authored a professional paper on the clouds, which continue to defy explanation. Although a team of astronomers have proposed that the plume was either a cloud of ice particles or a Martian aurora, neither theory fully explains the plume, raising new questions about the state of the Martian atmosphere.[2]

Before you dash out to buy a telescope to make your own discoveries, let us first review some helpful information and observing tips.

What Can You Expect to See?

Phases

Before and after opposition, and especially when Mars is either just beginning or near the end of its apparition, the three most obvious features we see are the planet's phase, colour and polar caps. At greatest phase, dark patches and Mars is nearly 90 per cent illuminated and resembles a gibbous Moon three days from full. The phase diminishes as the planet nears opposition (when it is full) and increases after opposition. We can never see Mars in a half or crescent, new or quarter phase, because the planet lies outside Earth's orbit.

Polar ice caps

Mars has two brilliant white polar ice caps that are the most noticeable features on the planet. As with Earth, Mars's polar axis is tilted with respect to its orbit around the Sun (25.2° compared to Earth's 23.5°), leading to seasons, during which we see the polar caps wax and wane in an annual cycle. Due to the axial tilt, we mostly see either one polar cap or the other but usually not both at the same time. Which cap we see when we look through our telescopes depends, once again, on the type of opposition: the south polar cap is tilted towards us during perihelic oppositions, while the north polar cap is tilted towards us at aphelic oppositions.

If we were to watch the south polar cap for an entire season, starting with the spring thaw, we would see the great ice sheet, which can extend halfway to the planet's equator, start to fracture

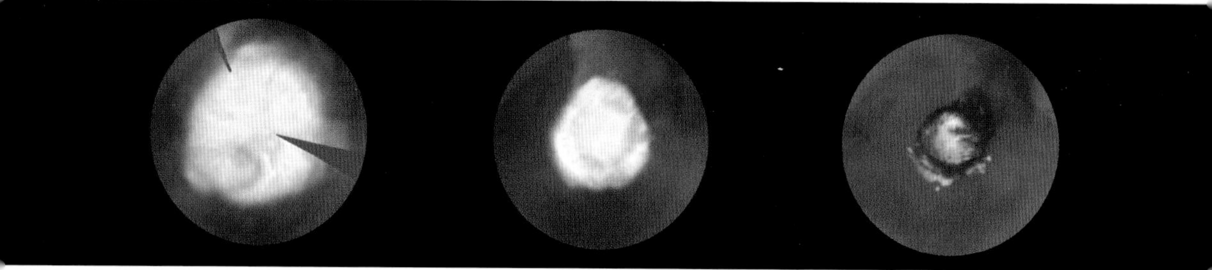

Three Hubble Space Telescope mosaic images showing the seasonal shrinking of Mars's North polar Cap, projected to make them appear as they would if seen from above. Left: October 1996, early Martian northern hemisphere spring, showing the cap's maximum extent. Middle: January 1997, mid-spring, showing the cap's average extent. Right: March 1997, early summer, showing the cap's minimum extent.

and shrink until, by the end of Martian summer, nothing remains of it but a tiny island of ice near the South Pole. The reason for this dramatic shift in appearance is that, unlike Earth's polar caps of water ice, those on Mars are a mix of water ice and carbon dioxide ice (a thick layer of water ice blanketed by carbon dioxide); so, as the Martian seasons change, the carbon dioxide ice sublimates (to pass directly from the solid to the vapour state) in summer, revealing the surface, before freezing again in winter.

Surface features

Next to the Martian phases and brilliant polar caps, the planet's dark surface patches (called albedo features) can be the most obvious features on the disc. The dark areas, once believed to be either bodies of water or tracts of vegetation, are now known to be areas of rocky outcrops that have been cleared of the ever shifting sands of Mars (the surrounding yellowish regions). Over time, the shape of the darkest regions can change as the winds blow and move the sands around. So consider the maps here as a foundation only.

We cannot see the entire disc at once. What regions we see depends on the date and time of our observation. Mars rotates on its axis with a period similar to Earth's: 24 hours 37 minutes; it also rotates in the same direction as Earth. So from hour to hour, its surface markings move from left to right over the course of the night as viewed through a simple inverting telescope (with south up and west to the left nearly 15° every hour). Also, because Mars rotates a bit more slowly than Earth, the features on Mars will appear about thirty minutes later each night, if we were to observe at the same time each night.

In a simple inverting telescope without a diagonal, and without a clock drive or tracking device, Mars will drift through the eyepiece from the right to the left toward the west. The side of Mars closest to the direction of motion is called the *preceding* limb. The trailing

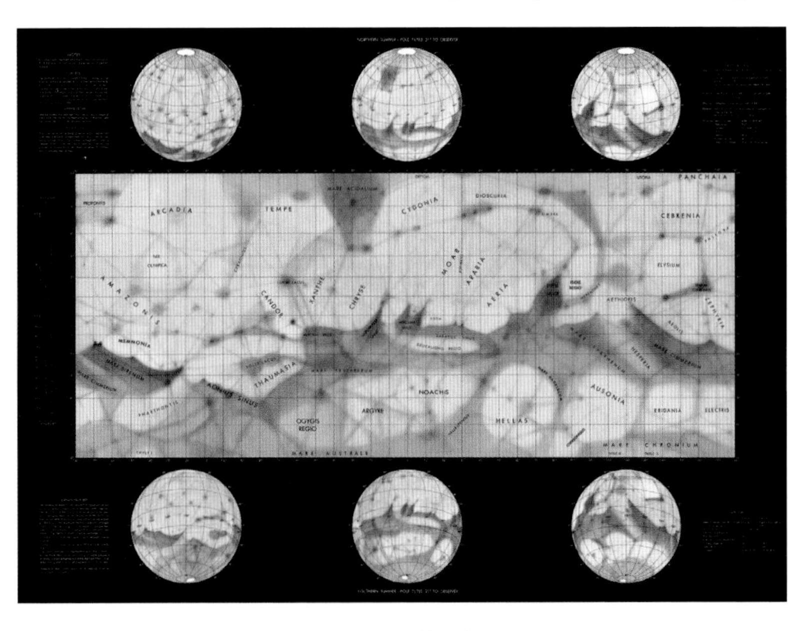

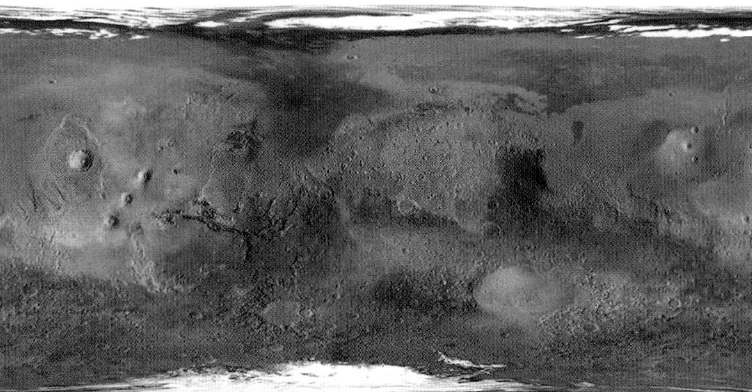

Top: A 1962 U.S. Airforce Aeronautical Chart and Information Center map of Mars showing the major named surface features of Mars (with labels) most likely to be visible through a backyard telescope of modest aperture (North is up and the following limb is to the left). As with Earth, Mars has north and south lines of latitude (from 0° at the equator, to -90° / +90° at the poles) and 360° longitude (measured westward from 0°). The darker the feature the more likely one will see it when that hemisphere is in view. Bottom: map of albedo feature superimposed on Martian surface details as recorded by the *Mars Global Surveyor* spacecraft. North is up.

limb (on the eastern side) is also called the *following* limb.

Note also that Mars independently spins on its axis over the course of a night, turning from the following limb (where it is sunrise on Mars; the *morning limb*) to the preceding limb (where it is sunset on Mars; the *evening limb*).

Let us look now at some of the most striking bright and dark features on Mars and what they represent. In these views, south is up to match the orientation in a simple inverting telescope. To identify these markings, first identify the nearest prominent feature described in this review, then refer to the map on the previous page to identify its Latin name.

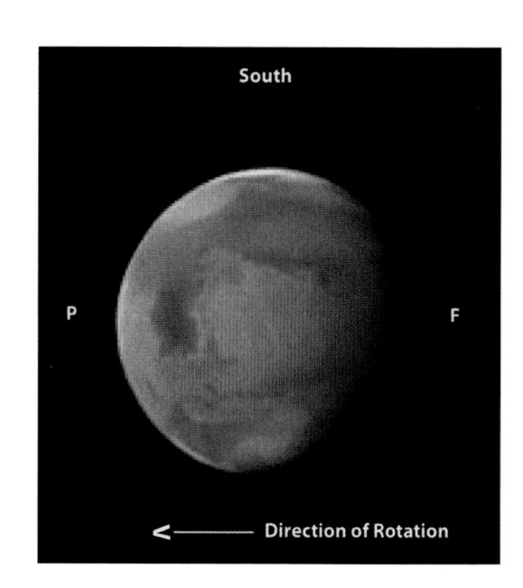

A simple inverted view of Mars (typical of that seen through a refractor or Newtonian reflector without a diagonal) showing the directions of the preceding (P) and following (F) limbs, as well as the direction in which the planet will rotate over the course of the night. As the Earth turns, the planet will drift through the telescope's field of view from the right to the left. Over the course of the night, Mars itself will also turn on its axis from the right to the left.

Syrtis Major (covering longitudes 270° to 315° W and latitudes 0° to 30° N) is prominently V-shaped and one of the darkest features on Mars and is therefore the most recognizable, especially during perihelic oppositions. This vast volcanic plateau's surface is as dark as fresh basalt flows on Earth or those of the lunar maria.

Hellas (also known as Hellas Planitia; centred roughly at 40° S, 290° W) lies just south of Syrtis Major. This bright and broad elevated ring surrounds a depression 7,000 kilometres (4,350 mi.) deep and 2,300 kilometres (1,430 mi.) wide. It is, in fact, one of the largest impact craters in the solar system, having been formed most likely by an asteroid when it collided with the planet early in its history, some 3.5 billion years ago.

Sinus Sabaeus-Meridiani are joint dark features, southwest of Syrtis Major, which run along the planet's equator: long Sinus Sabaeus to the east, and the forked plain Sinus Meridiani to the west. Sinus

Meridiani (meaning 'Meridian Bay') is the most prominent dark feature near 0° longitude and 0° latitude. Both Sinus Sabaeus and Sinus Meridiani are dark bedrock and fine-grained sand deposits ground down from ancient lava flows and other volcanic features. The western portion of Sinus Meridiani (known as Meridiani Planum) was the landing site of Mars Exploration Rover *Opportunity*.

Magaritifer Sinus A dark shark-tooth albedo feature, it lies west of Sinus Meridiani and is separated from it by a gulf of desert known as Aram. Its beak-like extension to the north sometimes appears broken off at the end. The Ares Valles, one of the largest Martian outflow channels, courses through the region's rough-and-tumble terrain, known as 'chaotic terrain'. Here we find heavily cratered highlands and massive impact basins that bear evidence of past water storage, subsequent drainage and massive flooding.

Mare Erythraeum The dusky region south of Magaritifer Sinus is occupied by Mare Erythraeum, whose boundaries are rather ill-defined. There is, however, one very notable feature, the large and bright circular formation Argyre (at ~50°S). This splendid feature, 1,800 kilometres (1,120 mi.) wide, is a 4-billion-year-old impact basin that may be one of the best places to search for life on Mars, since the impact may have helped to spread water and other life-supporting material from the crust to the surface, potentially creating a massive lake within the basin. Furthermore, when combined with activity from nearby volcanoes, energy from the impact could have helped to drive hydrothermal activity.[3]

Mare Acidaleum is a dark and expansive northern plain, one so large that it would cover a quarter of the continental United States on Earth. During aphelic oppositions, when Mars's north pole is tilted towards us, Mare Acidaleum is the most prominent feature on Mars. The dark butterfly marking just southeast of its borders, known as

Nilokeris, is noteworthy during some apparitions, but faint during others. The central part is characterized by a dark depression, which contains relatively smooth plains where several large outflow channels terminate. Earlier missions also revealed a striking dichotomy in the region: a heavily cratered plateau in the south that contrasts with low-lying, lightly cratered terrain in the north. In many places the boundary separating these two regions is a towering escarpment 1–2 kilometres (0.6–1.2 mi.) high.

Solis Lacus (also known as the Eye of Mars) is of great interest to Mars observers because it is a region famous for producing dust storms. As a result, the region is the most variable in appearance. The dark ocular lies within the centre of a vast high plain, ringed by mountains to the south and canyons to the north; the plateau sits at elevations from 2,500 metres (8,200 ft) to 4,300 metres (14,100 ft) above the surrounding terrain, and dozens of small shield volcanoes occupy the region's northwest sector. Continuous orbital monitoring by *Mars Global Surveyor*, and later by *Mars Reconnaissance Orbiter*, has shown that the region undergoes continual change, with drastic alterations during dust-storm seasons punctuating more subtle changes at other times of the year. The rapid surface changes are caused by the deposition and erosion of thin coatings of bright dust.[4]

Valles Marineris canyon system forms the northern brow of the Eye of Mars, just south of the equator. The largest canyon in the solar system, the valley extends more than 3,000 kilometres (1,865 mi.) in length, is 600 kilometres (375 mi.) across at its widest point and plunges 8 kilometres (5 mi.) deep. Geologists think Valles Marineris was created when earthquakes, generated by the growth of giant volcanoes nearby, opened a crack in the surface 3.5 billion years ago. As the crack widened, the ground sank, allowing subsurface water to escape and wash away fallen debis. At the valley's eastern end, scientists see unmistakable evidence for massive floods.

Landslides widened the canyon further until activity ceased about 2 billion years ago.

Tharsis The visual wonders of telescopic Mars are the mighty shield volcanoes of the Tharsis bulge, the largest volcanic region on Mars. It spans approximately 4,000 kilometres (2,485 mi.) across and is 10 kilometres (6 mi.) high, making them up to one hundred times larger than those anywhere on Earth. Although the bulge contains twelve large shield volcanoes, the three largest by far are *Arsia Mons*, *Pavonis Mons* and *Acraeus Mons*. *Olympus Mons* is on the edge.

Amazonis Planitia The northern hemisphere in this part of the planet is dominated by the light-coloured, relatively smooth to hummocky plains of Amazonis Planitia. The floor is comprised of a huge sheet of once liquid lava that had a crust of cooled debris floating on the surface. When the flow encountered some hills, the rubble crust was caught and piled up, forming thick masses of debris, which we see as a jumbled terrain. Downstream from the hills, there was no crust left and the lava formed a smoother, fresh surface.

Elysium supports the second-largest volcanic region on Mars. Lying atop a crustal rise that spans 1,700 by 2,400 kilometres (1,055 by 1,490 mi.) with numerous shield volcanoes, the site often contains bright orographic clouds, especially over its largest volcano (*Elysium Mons*), which measures 700 kilometres (435 mi.) across and rises 13 kilometres (8 mi.) above the surrounding plains. Near the planet's equator we find the flat smooth surface of the Elysium plains. This is the landing site for the *Interior Exploration using Seismic Investigations, Geodesy and Heat Transport* (InSight) spacecraft, designed to study the deep interior of Mars, which landed on 26 November 2018.

Mare Cimmerium and *Mare Tyrrhenum* are extensions of the dark southern highland feature Mare Sirenum. The entire region is heavily cratered terrain, which Antoniadi made famous when he described it as resembling the spots of a leopard. Like the lunar cratered highlands, those in the Martian southern highland region are ancient remnants of asteroid bombardment dating to more than 3.5 billion years ago. However, regions of Cimmerium contain gullies that may be due to recent flowing water; the region also is dappled with shifting sand dunes and shows evidence of glacial erosion. Mare Tyrrhenum also contains ancient cratered highlands and a volcanic complex that includes one of the oldest volcanoes on Mars (Tyrrhenus Mons). This region is as dark as neighbouring Hellas is bright, creating a wonderful visual contrast through backyard telescopes. The peninsula of Syrtis Major extends northward from the following edge of Mare Tyrrhenum.

We have now arrived at our final prominent feature in this tour of the Martian surface, which has taken us through a complete rotation. A slight turn of the globe will bring Syrtis Major back into the spotlight. Again, the features described here are only the darkest and brightest. On nights of excellent seeing, when the Red Planet is closest to Earth, you will certainly be able to identify other dimmer or less bright shades across the surface. If you are a first-time viewer of Mars, it is difficult to come away from the eyepiece with full assurance of all you have seen. But if the author's childhood memories hold true, even spying a dark feature like Syrtis Major for the first time can be a memorable moment. The beauty of observing today is that we can observe in the age-old traditions of the early Martian cartographers while enjoying the profound knowledge of these markings as defined to us by the many spacecraft now monitoring the planet on a daily basis. Such knowledge brings a new dimension to what the eye–brain system interprets. Whether it be the cobwebs of Lowell's canals or the leopard spots of Antoniadi's musings, the Red Planet can conjure up its own magic in our eyes.

Appendix I
Mars Oppositions, 2020–35

Opposition date	*Constellation*	*Apparent diameter*	*Distance from Earth (millions of km)*
13 October 2020	Pisces	22.3″	62.7
8 December 2022	Taurus	16.9″	82.3
16 January 2025	Gemini	14.4″	96.2
19 February 2027	Leo	13.8″	101.4
25 March 2029	Virgo	14.4″	97.1
4 May 2031	Libra	16.9″	83.6
27 June 2033	Sagittarius	22.0″	63.9
25 March 2035	Pisces	24.5″	57.1

Source: William Sheehan and Stephen James O'Meara, *Mars: The Lure of the Red Planet* (Amherst, NY, 2001)

APPENDIX II

MARS FACT SHEET

All data gleaned/adapted from NASA's Mars Fact Sheet. For additional data see https://nssdc.gsfc.nasa.gov/planetary/factsheet/marsfact.html

Mars/Earth comparison

Physical data

	Mars	Earth	Ratio
Equatorial diameter (km)	6,792	12,756	0.5
Core diameter (km)	3,400	6,970	0.5
Mass (10^{24} kg)	0.6	6.0	0.1
Mean density (kg/m^3)	3,933	5,514	0.7
Volume (10^{10} km^3)	16.3	108.3	0.15
Surface gravity (m/s^2)	3.7	9.8	0.4
Escape velocity (km/s)	5.0	11.2	0.45

Observational data

	Minimum	Maximum
Distance from Earth	(10^6 km) 55.7	(10^6 km) 401.3
Apparent diameter from Earth	3.5 (seconds of arc)	25.1
Apparent visual magnitude	+1.86	-2.94

Martian atmospheric data

Mean surface pressure (varies from 4.0 to 8.7 mb depending on season)	6.36 millibars (mb)	
Mean surface pressure	6.36 millibars (mb)	
Total mass of atmosphere	-2.5×10^{16} kg	
Average temperature	−63°C	
Daily temperature range (*Viking 1* site)	−31°C to −89°C	
Daily temperature range (*Curiosity* site)	−0°C to −70°C	
Minimum winter temperature at poles	−125°C	
Maximum summer temperature at equator	+20°C (during the day)	
Minimum summer temperature at equator	−73°C (during the night)	
Wind speeds (*Viking* lander sites)		
Summer	2–7 m/s	
Autumn	5–10 m/s	
Dust storms	17–30 m/s	
Composition (by volume)	*Mars*	*Earth*
Carbon dioxide (CO_2)	95.3%	0.04%
Nitrogen (N_2)	2.7%	2.7%
Argon (Ar)	1.6%	0.93%
Oxygen (O_2)	0.13%	20.95%
Water	210 (ppm)	0.1–4%

Orbital data

	Mars	Earth	Ratio
Orbital period (days)	6,867.0	365.2	1.9
Rotation period (hours)	24.7	24.0	1.03
Axial tilt (degrees)	25.2	23.4	1.1
Mean distance from Sun (10^6 km)	227.9	149.6	1.5
Perihelion distance (10^6 km)	206.6	147.1	1.4
Aphelion distance (10^6 km)	249.2	152.1	1.6
Orbit eccentricity	0.1	0.02	5.6
Orbital inclination (degrees)	1.85	0.0	—

Moons of Mars

	Phobos	Deimos
Mean orbital distance from centre of Mars	9,378	23,459
Orbital period (days)	0.3	1.3
Diameter (km)	22.2	12.6
Mass (10^{15} kg)	10.6	2.4
Surface temperature range	−4° to −112° C	~−113°C– −3°C
Density (gm/cm^3)	1.9	1.7
Orbital inclination (degrees)	1.1	1.8
Orbital eccentricity	0.015	0.0

Appendix III

Mars Missions

Mission	Country	Launch Date	Notes from NASA Space Science Data Coordinated Archive
Marsnik 1	USSR	10 Oct 1960	Launch failure
Marsnik 2	USSR	14 Oct 1960	Launch failure
Sputnik 22	USSR	24 Oct 1962	Failed to leave Earth orbit
Mars 1	USSR	1 Nov 1962	First spacecraft to fly by Mars (19 June 1963). Preceded by communication failure (21 March 1963)
Sputnik 24	USSR	4 Nov 1962	Failed to leave Earth orbit
Mariner 3	U.S.	5 Nov 1964	Failed soon after launch
Mariner 4	U.S.	28 Nov 1964	First successful flyby on 15 July 1965, takes first photos of Mars
Zond 2	USSR	30 Nov 1964	First successful firing of ion engines on an interplanetary mission. Communication lost prior to its Mars flyby on 6 August 1965
Mariner 6	U.S.	25 Feb 1969	Part of a successful dual mission with *Mariner 7*. *Mariner 6* returned 75 images, including close-ups that covered 20 per cent of Mars's surface
Mariner 7	U.S.	27 March 1969	Part of a successful dual mission with *Mariner 6*. *Mariner 7* returned 126 images including close-ups that covered 20 per cent of Mars's surface
Mars 1969A	USSR	27 March 1969	Launch failure
Mars 1969B	USSR	2 April 1969	Launch failure
Mariner 8	U.S.	9 May 1971	Launch failure
Cosmos 419	USSR	10 May 1971	Planned orbiter that failed in Earth orbit

Mars 2	USSR	19 May 1971	Successful arrival of orbiter on 27 November 1971. Descent module released that day but crashed onto surface
Mars 3	USSR	28 May 1971	Orbiter successfully arrives at Mars on 2 December 1971. Descent module successfully lands that day, but communications lost twenty seconds later, probably due to a massive dust storm. Transmission of vast amount of data from both *Mars 2* and *Mars 3* orbiters, and sixty images from landers, including those of towering Martian volcanoes
Mariner 9	U.S.	30 May 1971	Successful orbiter that arrived at Mars on 14 November 1971 and returned 7,329 images covering the entire planet, including the first detailed views of the Martian volcanoes, Valles Marineris, the polar caps, and the satellites Phobos and Deimos
Mars 4	USSR	21 July 1973	Orbiter reached Mars on 10 February 1974 but failed to achieve orbit due to an engine failure. The craft flew by the planet and returned one swathe of images and other data
Mars 5	USSR	25 Sept 1973	The spacecraft reached Mars on 12 February 1974 and collected data for 22 orbits until a pressurization failure ended the mission nine days later, but not before returning about sixty images
Mars 6	USSR	5 Aug 1973	Spacecraft reached Mars on 12 March 1974 and released a descent module that returned the first data from Mars's atmosphere for 224 seconds before transmission ceased
Mars 7	USSR	9 Aug 1973	Spacecraft reached Mars on 9 March 1974, prematurely releasing a lander that missed the planet
Viking 1	U.S.	20 Aug 1975	Orbiter achieved orbit on 19 July 1976, and successfully released a lander to the surface a day later, when transmission of the first surface images began 25 seconds after landing. The orbiter lasted four years and the lander operated for more than six
Viking 2	U.S.	9 Sept 1975	Orbiter achieved orbit on 7 August 1976, and successfully released a lander to the surface on 3 September. The orbiter lasted two years and the lander operated until 11 April 1980
Phobos 1	USSR	7 July 1988	Orbiter and lander lost on way to Mars due to a communication failure

Phobos 2	USSR	12 July 1988	Achieved Mars orbit but the mission ended prior to the craft reaching Phobos when the spacecraft signal was lost on 27 March 1989
Mars Observer	U.S.	25 Sept 1992	Contact with orbiter lost on 21 August 1993, three days before scheduled orbit insertion
Mars Global Surveyor	U.S.	7 Nov 1996	First spacecraft launched in a decade-long exploration of Mars by NASA. Orbiter achieved orbit on 12 September 1997 and operated for nine years and 52 days
Mars 96	Russia	16 Nov 1996	Failed to leave Earth orbit
Mars Pathfinder	U.S.	4 Dec 1996	Spacecraft successfully landed on Mars on 4 July 1997 and renamed Carl Sagan Memorial Station. Its *Sojourner* rover was successfully deployed onto the surface, becoming the first to operate outside the Earth–Moon system. Images were taken and experiments performed by the lander and rover until 27 September 1997, when communications were lost for unknown reasons
Nozomi	Japan	3 July 1998	Japan's first Mars explorer failed to reach the planet due to an electrical failure
Mars Climate Orbiter	U.S.	11 Dec 1998	Reached Mars on 23 September 1999 but radio contact was lost shortly thereafter
Mars Polar Lander	U.S.	3 Jan 1999	Spacecraft entered Mars's atmosphere on 3 December 1999 but signal was lost prior to landing
Deep Space 2	U.S.	3 Jan 1999	Two probes attached to *Mars Polar Lander*, lost during failed mission
Mars Odyssey	U.S.	4 July 2001	Orbiter reached Mars on 24 October 2001 and continues to collect data from Mars orbit
Mars Express	ESA	2 June 2003	First ESA planetary mission successfully achieves Mars orbit on 25 December 2003 and continues to collect data
Beagle 2	ESA	2 June 2003	Six days prior to its Mars orbit insertion, *Mars Express* released its *Beagle 2* lander, which was lost prior to landing
Spirit	U.S.	10 June 2003	Rover successfully landed on Mars on 4 January 2004. The last transmission occurred on 22 March 2010. The rover travelled a total of 7.73 kilometres over a period of six years, two months

Opportunity	U.S.	8 July 2003	The Rover successfully landed on Mars on 25 January 2004 and continued to operate until 10 June 2018, when transmission failed after a global dust storm. Mission officially ended February 2019
Mars Reconnaissance Orbiter	U.S.	12 Aug 2005	Spacecraft successfully achieved Mars orbit on 10 March 2006 and continues to collect data from orbit
Phoenix	U.S.	4 Aug 2007	Lander touched down successfully on 25 May 2008, farther north than any previous spacecraft sent to Mars. It returned more than 25,000 pictures before the mission ended five and a half months later
Phobos-Grunt	Russia	8 Nov 2011	Failed to leave Earth orbit
Yinghuo-1	China	8 Nov 2011	Launched piggyback on the doomed Russian Phobos-Grunt spacecraft, the mission ended in failure
Curiosity	U.S.	26 Nov 2011	Lander successfully touched down on Mars on 5 August 2012 and continues to roam the Martian surface
Mars Orbiter Mission	India	5 Nov 2013	Formerly Mangalyaan, the spacecraft successfully achieved Mars orbit on 21 September 2014 and continues to collect data from orbit – though the designed mission life was only six months
MAVEN	U.S.	18 Nov 2013	Orbiter successfully achieved Mars orbit on 22 September 2014 and continues to collect data from orbit on the planet's climate history
ExoMars	ESA	14 March 2016	Successfully achieved orbit on 19 October 2016 and continues to collect data from orbit

References

1 Communion with Mars

1 Antonin Pannekoek, *A History of Astronomy* (New York, 1961), pp. 39–40.
2 Yngve Vogt, 'World's Oldest Ritual Discovered: Worshipped the Python 70,000 Years Ago', *Apollon*, 1 February 2012, www.apollon.uio.no.
3 Stephen James O'Meara, *Night Skies of Botswana: Suitable for all Stargazing in the Southern Hemisphere* (Cape Town, 2019).
4 Stephen James O'Meara, private communication with local people, 30 July 2018.
5 Personal observation and private communication; see also Elana Bregin, 'The Identity of Difference: A Critical Study of Representations of the Bushmen', partial submission for award of MA, University of Natal, 1998.
6 Stephen James O'Meara, private communication with Ghanzi Basarwa, 17 April 2017.
7 Robert Stevens Fuller, 'The Astronomy of the Kamilaroi and Euahlayi Peoples and Their Neighbours', MPhil thesis, Macquarie University, Sydney, 2014.
8 Stephen James O'Meara, 'Tales from the Pacific', *Sky and Telescope*, LXXII/73 (1986), p. 73; also Stephen Robert Chadwick and Martin Paiour-Smith, *The Great Canoes in the Sky: Starlore and Astronomy of the South Pacific* (New York and London, 2017).
9 Duane W. Hamacher, 'Observations of Red-giant Variable Stars by Aboriginal Australians', *Australian Journal of Anthropology*, XXIX (2018), pp. 89–107.
10 *The Concise Mythological Dictionary* (London, 1963) p. 115; and Richard Olson, *Science Deified and Science Defied: The Historical Significance of Science in Western Culture* (Berkeley, CA, 1982), p. 44.
11 Nicholas Campion, *Astrology and Cosmology in the World's Religions* (New York and London, 2012), p. 132.
12 Enn Kasak and Raul Veede, 'Understanding Planets in Ancient Mesopotamia', *Folklore*, 16 (2001), pp. 6–33, available at www.folklore.ee.

13 R. A. Parker, 'Ancient Egyptian Astronomy', *Philosophical Transactions of the Royal Society of London*, CCLXXVI/1257 (1974), pp. 51–65.
14 Bojan Novaković, 'Senenmut: An Ancient Egyptian Astronomer', *Publications of the Astronomical Observatory of Belgrade*, 85 (2008), pp. 19–23.
15 Parker, 'Ancient Egyptian Astronomy'.
16 Leonard Schmitz, 'Mars', in *Dictionary of Greek and Roman Biography and Mythology*, ed. William Smith, 3 vols (London, 1846), vol. II, p. 961.
17 Ibid.
18 Alexander Jones, ed., *Ptolemy in Perspective: Use and Criticism of his Work from Antiquity to the Nineteenth Century* (New York, 2010).
19 'Commentariolus', from Edward Rosen's Introduction to *Nicholas Copernicus: Minor Works* (Warsaw and Cracow, 1985), https://copernicus.torun.pl.
20 Charles Nevers Holmes, 'Nicolaus Copernicus', *Popular Astronomy*, vol. XXIV (1916), p. 218.
21 Stephen Sols, 'Copernicus and the Church: What the History Books Don't Say', *Christian Science Monitor*, 19 February 2013, www.csmonitor.com.
22 Andreas Kleinert quoted in Michael Schaff, 'Dispelling Myths and Highlighting History of the Heliocentric Model', *Physics Today*, LXI (June 2008), p. 10.
23 R. A. Gray, 'The Life and Work of Tycho Brahe', *Journal of the Royal Astronomical Society of Canada*, XVII (1923), p. 104; and Bernard le Bovier de Fontenelle, *Conversations on the Plularity of Worlds*, trans. H. A. Hargreaves (Berkeley and Los Angeles, CA, and London, 1990), p. 56.
24 Owen Gingerich, 'Johannes Kepler and the New Astronomy', *Quarterly Journal of the Royal Astronomical Society*, XIII (1972), pp. 346–73.
25 Ibid.

2 The 'Miniature of Our Earth'

1 Galileo Galilei and Johannes Kepler, *The Sidereal Messenger of Galileo Galilei and a Part of the Preface to Kepler's Dioptrics, Containing the Original Account of Galileo's Astronomical Discoveries*, ed. and trans. Edward Stafford Carlos (London, 1880), p. 109.
2 Galileo Galilei, *Le Opere di Galileo Galilei*, ed. Antonio Favaro (Florence, 1890), vol. X, p. 503.
3 Paolo Molaro, 'Francesco Fontana and his *Astronomical* Telescope', *Journal of Astronomical History and Heritage*, XX/2 (2017), pp. 271–88.
4 Camille Flammarion, *Camille Flammarion's The Planet Mars*, trans. Patrick Moore, ed. William Sheehan (New York and London, 2015), p. 14.
5 Ibid., p. 15.
6 Ibid., p. 31.

7 Richard McKim, 'Telescopic Martian Dust Storms: A Narrative and Catalogue', *Memoirs of the British Astronomical Association*, XLIV (1999), pp. 14–15.
8 Percival Lowell, 'The Polar Snows', *Popular Astronomy*, II (1894), p. 52.
9 William Herschel, 'On the Remarkable Appearances at the Polar Regions of the Planet Mars, the Inclination of its Axis, the Position of its Poles, and its Spheroidal Figure; with a Few Hints Relating to its real Diameter and Atmosphere', in *The Scientific Papers of Sir William Herschel* (London, 1912), vol. I, p. 156.
10 Richard McKim, 'The Dust Storms of Mars', *Journal of the British Astronomical Association*, CVI (1996), p. 186.
11 Flammarion cited in Thomas Hockey with Virginia Trimble and Thomas R. Williams, eds, *Biographical Encyclopedia of Astronomers* (New York and London, 2007), p. 724.
12 Camille Flammarion, 'Mars, by the Latest Observations', *Popular Science*, IV (1873), p. 179 and p. 190.
13 Richard A. Proctor, *Other Worlds Than Our Own* (New York, 1871), p. 100.
14 William Huggins, 'On the Spectrum of Mars, with Some Remarks on the Colour of the Planet', *Monthly Notices of the Royal Astronomical Society*, XXVII (1867), p. 179 and p. 190.
15 William Sheehan, *Planets and Perception: Telescopic Views and Interpretations, 1609–1908* (Tucson, AZ, 1988), p. 179 and p. 191.
16 Percival Lowell, *Mars* (Boston, MA, and New York, 1895), p. 119.
17 David Sutton Dolan, 'Percival Lowell: The Sage as Astronomer', PhD diss., University of Wollongong, 1992, p. 72.
18 'Impossibility of the Linear "Canal" Network of Schiaparelli as an Objective Reality on the Planet', Report of the Mars Section, *Memoirs of the British Astronomical Association*, XX (1916), pp. 38–45.
19 E. M. Antoniadi, *The Planet Mars*, trans. Patrick Moore (Chatham, 1975), p. 33.
20 Robert Crossley, 'Percival Lowell and the History of Mars', *Massachusetts Review*, XLI/3 (2000), pp. 297–318.

3 Romancing Mars

1 Robert Crossley, 'Mars and the Paranormal', *Science Fiction Studies*, XXXV/3 (2008), pp. 466–84.
2 Ibid.
3 Théodore Flournoy, *From India to the Planet Mars: A Study of a Case of Somnambulism* (New York and London, 1900), p. 234.
4 H. G. Wells, *The War of the Worlds* (London, 1898), p. 3.

5 Robert Markley, *Dying Planet: Mars in Science and the Imagination* (Durham, NC, 2005), p. 25.
6 William Sheehan and Stephen James O'Meara, *Mars: The Lure of the Red Planet* (Amherst, NY, 2001), p. 195.
7 Hadley Cantril, *The Invasion from Mars: A Study in the Psychology of Panic* (Piscataway, NJ, 1940), p. ix.
8 Ibid., p. 68.
9 Richard S. Lewis, *From Vinland to Mars: A Thousand Years of Exploration* (New York, 1978), p. 356.
10 William Huggins, 'On the Spectrum of Mars, with Some Remarks on the Colour of the Planet', *Monthly Notices of the Royal Astronomical Society*, XXVII (1867), p. 180.
11 David H. De Vorkin, 'W. W. Campbell's Spectroscopic Study of the Martian Atmosphere', *Quarterly Journal of the Royal Astronomical Society*, XVIII (1977), p. 38.
12 Françoise Launay, *The Astronomer Jules Janssen: A Globetrotter of Celestial Physics*, trans. Storm Dunlop (New York and London, 2012), p. 31.
13 W. H. Pickering to E. C. Pickering, Harvard College Observatory Director's correspondence (Cambridge, MA, 1877).
14 Howard Plotkin, 'William H. Pickering in Jamaica: Founding of Woodlawn and Studies of Mars', *Journal for the History of Astronomy*, XXI (1993), p. 101.
15 W. W. Campbell, 'The Spectrum of Mars', *Publications of the Astronomical Society of the Pacific*, VI (1894), p. 228.
16 W. S. Adams and C. E. St John, 'An Attempt to Detect Water-vapor and Oxygen Lines in the Spectrum of Mars with the Registering Microphotometer', *Publications of the Astronomical Society of the Pacific*, XXXVII/37 (1926), pp. 158–9.
17 De Vorkin, 'W. W. Campbell's Spectroscopic Study of the Martian Atmosphere', p. 50.
18 Alfred R. Wallace, *Man's Place in the Universe: A Study of the Results of Scientific Research in Relation to the Unity or Plurality of Worlds* (London, 1904), p. 97.
19 Carl Sagan, Tobias Owen and H. J. Smith, *Planetary Atmospheres* (Dordrecht, 1971), p. 225.
20 R. W. Buchheim, *Space Handbook: Astronautics and Its Applications: 85th Congress, 2nd session, House of Representatives staff report* (1960); and W. M. Sinton, 'Further Evidence of Vegetation on Mars', presented at the meeting of the American Astronomical Society, Gainesville, FL, 27–30 December 1958, https://history.nasa.gov.

4 The First Emissaries to Mars

1 Quoted in Roger D. Launius, 'Sputnik and the Origins of the Space Age', NASA, https://history.nasa.gov, accessed 20 February 2019.
2 Ibid.
3 Suzanne Deffree, '1st U.S. Satellite Attempt Fails, December 6, 1957', 6 December 2018, www.edn.com.
4 'Explorer 1 Overview', NASA, www.nasa.gov, accessed 20 February 2019.
5 'Marsnik 1', NASA, https://nssdc.gsfc.nasa.gov, accessed 26 February 2018.
6 'Sputnik 22', NASA, https://nssdc.gsfc.nasa.gov, accessed 20 February 2019.
7 'Mars 1', NASA, https://nssdc.gsfc.nasa.gov, accessed 27 February 2018.
8 'Mariner 3', NASA, https://nssdc.gsfc.nasa.gov, accessed 27 February 2018.
9 Tim Wallace, 'First Mission to Mars: Mariner 4's Special Place in History', *Cosmos Magazine*, 14 July 2017, https://cosmosmagazine.com.
10 'Report from Mars: Mariner IV, 1964–1965', NASA, www.scribd.com, accessed 1 February 2019.
11 'The Dead Planet', *New York Times*, 30 July 1965, as reported in 'On Mars: Exploration of the Red Planet, 1958–1978', www.hq.nasa.gov, accessed 28 February 2018.
12 'Mariner 9', NASA, https://nssdc.gsfc.nasa.gov, accessed 1 March 2018.
13 Richard McKim, 'The Dust Storms of Mars', *Journal of the British Astronomical Association*, CVI (1996), p. 185.
14 'Mariner 9 Anniversary/Landslides on Mars', NASA, www.jpl.nasa.gov, accessed 1 March 2018.
15 'Missions to Mars', The Planetary Society, www.planetary.org, accessed 1 March 2018.
16 Quoted in Samuel Glasstone, *The Book of Mars* (Washington, DC, 1968), p. 217.
17 Ibid.
18 Carl Sagan, 'The Solar System', *Scientific American*, CCXXXIII (1975), pp. 22–31.
19 Quoted in Joe Achenbach, 'NASA's 1976 Viking Mission to Mars did all that was Hoped for – Except Find Martians', *Washington Post*, 18 June 2016, www.adn.com.
20 'Viking Mission to Mars', NASA, www.jpl.nasa.gov, accessed 20 February 2019.
21 Carl Sagan quoted in David McNab and James Younger, *The Planets* (New Haven, CT, 1999), p. 193.
22 'Searching for Life on Mars: Development of the Viking Gas Chromatograph Mass Spectrometer', NASA, https://appel.nasa.gov, accessed 20 February 2019.

23 Ibid.
24 Stewart Brand, 'Controversy is Rife on Mars: Interviewing Carl Sagan and Lynn Margulis', National Space Society, https://space.nss.org, accessed 20 February 2019.
25 'Viking 1 and 2', NASA, https://mars.nasa.gov, accessed 12 March 2018.
26 Brand, 'Controversy is Rife on Mars'.
27 Ibid.
28 Ibid.
29 'Viking Mission to Mars', NASA, www.jpl.nasa.gov, accessed 20 February 2019.
30 'Project Viking Fact Sheet', Jet Propulsion Laboratory, https://astro.if.ufrgs.br, accessed 12 March 2018.
31 Erik Conway, 'The 1990s (Overview)', JPL History, www.jpl.nasa.gov, accessed 30 September 2019.

5 Lifting the Curse at Mars

1 Douglas Messier, 'Will Russia End its Curse at Mars?', *The Space Review*, 7 November 2011, www.thespacereview.com.
2 Edwin V. Bell, II, 'Phobos Project Information', NASA (2016), https://nssdc.gsfc.nasa.gov, accessed 1 March 2019.
3 'Mars-96', www.russianspaceweb.com, accessed 1 March 2019.
4 Erik Conway, 'The 1990s (Overview)', JPL History, www.jpl.nasa.gov, accessed 30 September 2019.
5 Quoted in Diane Ainsworth, 'Pathfinder Results Featured in This Week's Science Magazine', NASA, https://mars.nasa.gov, accessed 21 March 2018.
6 'Mars Pathfinder', NASA, https://mars.nasa.gov, accessed 21 March 2018.
7 'Mars Pathfinder Winds Down After Phenomenal Mission', NASA, https://mars.jpl.nasa.gov, accessed 12 March 2018.
8 Susannah Fox, 'The Internet Circa 1998', The Pew Research Center, 21 June 2007, www.pewinternet.org.
9 'Home Computer Use and the Internet', Child Trends Databank, www.childtrends.org, 2015, accessed 1 March 2019.
10 Naona Yates, 'Millions Visit Mars – on the Internet', *Los Angeles Times*, 14 July 1997, https://articles.latimes.com.
11 Brian Dunbar, 'The Day the Internet Stood Still', NASA, www.nasa.gov, accessed 12 March 2018.
12 'Mars Global Surveyor: Important Discoveries', NASA, https://mars.jpl.nasa.gov, accessed 13 September 2019.
13 Paul Geissler, 'Three Decades of Martian Surface Changes', *Journal of Geophysical Research*, CX/E2 (2005).

14 Guy Webster, 'NASA – Report Reveals Likely Causes of Mars Spacecraft Loss', NASA, 13 April 2007, www.nasa.gov.
15 'Deep Space 2 Probes Silent', 6 December 1999, NASA, https://mars.nasa.gov.
16 'Mars Polar Lander/Deep Space 2: About the Mission', NASA, www.jpl.nasa.gov, accessed 23 September 2019.
17 P. R. Christensen et al., 'Evidence for Magmatic Evolution and Diversity on Mars from Infrared Observations', *Nature*, 436 (28 July 2005), pp. 504–9.
18 'Found It! Ice on Mars!', NASA, https://science.nasa.gov, accessed 14 March 2018.
19 Guy Webster, 'Mars May Be Emerging from an Ice Age', Jet Propulsion Laboratory, www.jpl.nasa.gov, accessed 14 March 2018.
20 'Mars Express Science Highlights: #4. Probing the Polar Regions', ESA, 3 June 2013.
21 'Mars Express Confirms Methane in the Martian Atmosphere', ESA, 30 March 2004, www.esa.int.
22 'Buried Glaciers on Mars', ESA, www.esa.int, accessed 22 March 2018.
23 'First Mars Express Gravity Results Plot Volcanic History', ESA, www.esa.int, accessed 22 March 2018.
24 Markus Bauer, 'Shining Light on the Aurora of Mars', ESA, www.esa.int, accessed 8 March 2018.
25 Markus Bauer, 'Leaky Atmosphere Linked to Lightweight Planet', ESA, 8 February 2018, https://sci.esa.int.
26 'Spirit and Opportunity', NASA, https://mars.nasa.gov, accessed 15 March 2018.

6 Scouting Out Humanity's Next Home

1 Quoted in 'Mars Reconnaissance Orbiter: NASA Science Goals', Jet Propulsion Laboratory (JPL), www.nasa.gov, accessed 20 February 2019.
2 Quoted in Guy Webster, Jennifer LaVista and Laurie Cantillo, 'Steep Slopes on Mars Reveal Structure of Buried Ice', NASA, 11 January 2018, https://mars.nasa.gov.
3 Quoted in Nola Taylor Redd, 'Mars Volcanic Glass May Be Hotspot for Life', NBC News, 1 May 2012, www.nbcnews.com.
4 'Gemstone of the Year – Labeled', NASA, 28 October 2008, https://mars.nasa.gov.
5 Preston Dyches, 'NASA Radar Finds Ice Age Record in Mars' Polar Cap', JPL, 26 May 2016, www.jpl.nasa.gov.
6 Guy Webster, 'NASA Spacecraft Confirms Martian Water, Mission Extended', NASA, www.nasa.gov, accessed 22 September 2019.

7 Dwayne Brown et al., 'Mars Phoenix Lander Finishes Successful Work on Red Planet', NASA, 10 November 2008, www.nasa.gov.
8 Tereza Pultarova, 'Phobos-Grunt Probe Adds Chapter to Russian Space Disaster Series', *Space Safety Magazine*, 11 November 2011, www.spacesafetymagazine.com.
9 Emily Lakdawalla, 'Curiosity's Discoveries on Mars', *Sky and Telescope* (April 2017), p. 14.
10 'Curiosity Mission Overview', NASA/Jet Propulsion Laboratory, https://mars.nasa.gov, accessed 1 December 2018.
11 Quoted in Alan Boyle, 'Curiosity Rover Sees Life-friendly Conditions in Ancient Mars Rock', NBC News, 12 March 2013, www.nbcnews.com.
12 Eric Hand, 'On Mars, Atmospheric Methane – A Sign of Life on Earth – Changes Mysteriously with the Seasons', *Science*, 3 January 2018, www.sciencemag.org.
13 Quoted ibid.
14 'Mars Water-ice Clouds Are Key to Odd Thermal Rhythm', NASA, www.nasa.gov, accessed 22 September 2019.
15 'NASA's MAVEN Reveals Most of Mars' Atmosphere Was Lost to Space', NASA, 30 March 2017, www.nasa.gov.
16 Ibid.
17 Oleg Abramov and Stephen J. Mojzsis, 'Thermal Effects of Impact Bombardments on Noachian Mars', *Earth and Planetary Science Letters*, 442 (15 May 2016), pp. 108–20.
18 Stephen Mojzsis, quoted in 'Early Mars Bombardment Likely Enhanced Life-supporting Habitat', University of Colorado Press Release, www.colorado.edu, accessed 23 September 2019.
19 'Mars Orbiter Mission (MOM) Completes 4 Years in its Orbit', Indian Space Research Organisation, 24 September 2018, www.isro.gov.in
20 Rich Zurek, quoted in 'NASA encounters the Perfect Storm for Science', University of Colorado Boulder, http://lasp.colorado.edu, accessed 23 September 2019.
21 'Mars Dust Storm', ESA, 19 July 2018, www.esa.int.
22 Tony Greicius, ed., 'What Happened to Early Mars' Atmosphere? New Study Eliminates One Theory', NASA, 8 August 2017, www.nasa.gov.
23 'NASA Encounters the Perfect Storm for Science', NASA, 18 June 2018, https://mars.nasa.gov.
24 'ExoMars Highlights Radiation Risk for Mars Astronauts, and Watches as Dust Storm Subsides', https://exploration.esa.int, 19 September 2018.
25 'Mars Express Detects Liquid Water Hidden Under Planet's South Pole', ESA, 25 July 2018, www.esa.int.
26 'ExoMars Highlights Radiation Risk for Mars Astronauts'.

27 Ibid.
28 'Radiation Exposure Comparisons with Mars Trip Calculation', NASA, 1 January 2019, www.nasa.gov.
29 Trent Perrotto and Deb Schmid, 'Radiation Measured by NASA's Curiosity on Voyage to Mars has Implications for Future Human Missions', NASA, 30 May 2013, www.nasa.gov.
30 Avi Selk, 'Stephen Hawking, Kris Jenner and Other Famous People with Plans to Send Humans to Mars', *Washington Post*, 22 June 2017, www.washingtonpost.com.
31 Ben Guarno, 'Stephen Hawking Calls for a Return to the Moon as Earth's Clock Runs Out', *Washington Post*, 21 June 2017, www.washingtonpost.com.
32 'Mars Terraforming Not Possible Using Present-day Technology', NASA, 30 July 2018, www.nasa.gov.

7 Mars: Our Home Away from Home?

1 Eli Rosenberg, 'In the Coldest Village on Earth, Eyelashes Freeze, Dinner is Frozen and Temperatures Sink to –88°F', *Washington Post*, 16 January 2018.
2 'Hawai'i Space Exploration Analog and Simulation', Hi-Seas, https://hi-seas.org, accessed 25 January 2018.
3 Claude A. Piantadosi, *The Biology of Human Survival: Life and Death in Extreme Environments* (Oxford, 2003), p. 230.
4 Jacqueline Yallop, 'Force of Nature: Picturing Ruskin's Landscape', www.guildofstgeorge.org.uk, accessed 21 January 2018.
5 'Journey to Mars Overview', NASA, www.nasa.gov, accessed 25 January 2018.
6 'Mars Base Camp', Lockheed Martin, https://lockheedmartin.com, accessed 22 January 2018.
7 T. S. Eliot, Preface to Harry Crosby, *Transit of Venus: Poems* (Paris, 1931).
8 Sarah Frazier, 'Real Martians: How to Protect Astronauts from Space Radiation on Mars', NASA, 7 August 2017, www.nasa.gov.
9 Rashmi Shivni, 'NASA Twins Study Spots Thousands of Genes Toggling On and Off in Scott Kelly', PBS, 28 October 2017, www.pbs.org.
10 Ibid.
11 Amy Blanchett, 'Firworks in Space: NASA's Twin Study Explores Genes Expression', NASA, www.nasa.gov, accessed 22 September 2019.
12 S. M. Smith et al., 'Benefits for Bone from Resistance Exercise and Nutrition in Long-duration Spaceflight: Evidence from Biochemistry and Densitometry', *Journal of Bone and Mineral Research*, XXVII/9 (2012), pp. 1896–1906.
13 Kathryn Powley, 'Getting Sick in Space on the Way to Mars', University of Melbourne, 5 October 2017, https://pursuit.unimelb.edu.au.

14 Mike Wall, 'Hibernating Astronauts May Be Key to Mars Colonization', *Space.com*, 30 August 2016, www.space.com.
15 Andy Weir, 'If We're Serious about Going to Mars, We Need Artificial Gravity', 4 March 2014, www.space.com.
16 'Space Launch System: Building America's New Rocket for Deep Space Exploration', NASA, www.nasa.gov, accessed 20 January 2018.
17 'Lockheed Martin Reveals New Details to its Mars Base Camp Vision', Lockheed Martin, 28 September 2017, https://news.lockheedmartin.com.
18 Mary Harden, 'Odyssey Finds Water Ice in Abundance Under Mars' Surface', Jet Propulsion Laboratory, 28 May 2002, www.jpl.nasa.gov.
19 Guy Webster, 'NASA Phoenix Mars Lander Confirms Frozen Water', NASA, 20 June 2008, www.nasa.gov.
20 Colin M. Dundas et al., 'Exposed Subsurface Ice Sheets in the Martian Mid-latitudes', *Science*, 359 (12 January 2018), pp. 199–201.
21 Mallory Locklear, 'NASA Tests Small Nuclear Reactor that Could Power a Habitat on Mars', NASA, 18 January 2018, www.engadget.com.
22 Justin Joffe, 'NASA's Kilopower Project Will Bring Us to Mars and Beyond With Nuclear Power', *Outer Places*, 22 January 2018, www.outerplaces.com.
23 'A New Home on Mars: NASA Langley's Icy Concept for Living on the Red Planet', NASA, 29 December 2016, www.nasa.gov.
24 'Mars Ice House', www.marsicehouse.com, accessed 25 January 2018.
25 Lacey Cooke, 'NASA is Sending its Inflatable Mars Ice Home into Space', NASA, 25 October 2017, https://inhabitat.com.
26 'Human Needs: Sustaining Life During Exploration', NASA, 16 April 2007, www.nasa.gov.
27 Mark Garcia, ed., 'Space Food Bars Will Keep Orion Weight Off and Crew Weight On', NASA, 22 November 2016, www.nasa.gov.
28 Mike Wall, 'Mars Soil May Be Toxic to Microbes', *Space.com*, 6 July 2017, www.space.com.
29 Cyril Verseux et al., 'Sustainable Life Support on Mars – The Potential Roles of Cyanobacteria', *International Journal of Biology*, XV (2016), p. 65.
30 'Melissa', ESA, www.esa.int, accessed 28 January 2018.
31 'Waste Not, Want Not, On The Road To Mars', ESA, www.esa.int, accessed 1 February 2018.
32 'Earthworms Can Reproduce in Mars Soil Simulant', Wageningen University & Research, www.wur.nl, accessed 1 February 2018.
33 'Recycling for Moon, Mars and Beyond', NASA, www.nasa.gov, accessed 31 January 2018.
34 Dave Mosher, 'Elon Musk says SpaceX is on Track to Launch People to Mars within 6 Years: Here's the Full Timeline of his Plans

to Populate the Red Planet', *Business Insider*, 2 November 2018, www.businessinsider.com.

35 Michael Meltzer, 'When Biospheres Collide: A History of NASA's Planetary Protection Programs', NASA, Publication SP-2011-4234 (2010).

36 Aaron Gronstal, 'Putting the Ethics into Planetary Protection', NASA, 13 August 2018, https://astrobiology.nasa.gov.

37 Meltzer, 'When Biospheres Collide'.

8 The Lilliputian Moons of Mars

1 Stephen James O'Meara, 'The Demon Sprites of Mars', *Sky and Telescope*, CI/6 (2001), p. 102.

2 Asaph Hall, 'The Discovery of the Satellites of Mars', *Monthly Notices of the Royal Astronomical Society*, XXXVIII (1878), p. 205.

3 Angelo Hall, *An Astronomer's Wife: A Biography of Angeline Hall* (Baltimore, MD, 1908).

4 Robert Richardson, *Exploring Mars* (New York, 1954), p. 162.

5 Angelo Hall, 'Epilogue', *An Astronomer's Wife*, p. 130.

6 William Sheehan, *The Planet Mars* (Tucson, AZ, 1966), p. 63.

7 John Westfall and William Sheehan, *Celestial Shadows: Eclipses, Transits, and Occultations* (New York, 2015), p. 234.

8 Jonathan Swift, *Gulliver's Travels* [1726] (Philadelphia, PA, and New York, 1800), p. 219.

9 Owen Gingerich, 'The Satellites of Mars: Prediction and Discovery', *Journal for the History of Astronomy*, I (1970), pp. 109–15.

10 Ibid.

11 V. M. Zharkov and A. V. Kozenko, 'Fobos I Deimos – sputniki Marsa', *Novoye v zhizni, nauke, tekhnike, seriya: kosmonavtika, astronomiya*, I (Moscow, 1985), pp. 1–64; trans. as 'Phobos and Deimos: Satellites of Mars', NASA Technical Memorandum TM-88016, www.researchgate.net, accessed 7 May 2019.

12 S. Kostinsky, 'Sur les photographies des satellites de Mars', *Astronomische Nachrichten*, CLXXXVI (1909), p. 7.

13 Zharkov and Kozenko, 'Phobos and Deimos'.

14 E. C. Pickering, A. Searle and W. Upton, 'Photometric Observations: The Satellites of Mars', *Annals of the Harvard College Observatory*, XI (1879), pp. 226–38.

15 D. Pascu, S. Erard, W. Thuillot and V. Lainey, 'History of Telescopic Observations of the Martian Satellites', *Planetary and Space Science*, CII (2014), pp. 2–8.

16 Brad Smith, 'Phobos Preliminary Results from Mariner 7', *Science*, CLXVIII (1970), pp. 828–30.

17 Voltaire, *Micromégas*, https://ebooks.adelaide.edu.au, accessed 22 September 2019.
18 'Martian Moons', NASA, https://history.nasa.gov, accessed 19 February 2018.
19 'The Complete Phobos 2 VSK Data Set', The Planetary Society, www.planetary.org, accessed 18 February 2018.
20 'Phobos', JPL, https://mars.jpl.nasa.gov, accessed 20 February 2019.
21 T. D. Glotch et al., 'Spectral Properties of Phobos from the Mars Global Surveyor Thermal Emission Spectrometer: Evidence for Water and Carbonate', 46th Lunar and Planetary Science Conference: The Woodlands, TX, 2015, p. 2587.
22 Guy Webster, 'NASA Spacecraft Images Mars Moon in Color and in 3D', NASA, 14 July 2009, www.nasa.gov.
23 See 'Deimos as Viewed by Mars Reconnaissance Orbiter', The Planetary Society, www.planetary.org, accessed 20 February 2018.
24 Henning Krause and Ralf Jaumann, 'Images of Mars Express' Closest Ever Flyby at Phobos', ESA, 30 July 2008, www.dlr.de.
25 'Martian Moons: Phobos', ESA, https://sci.esa.int, accessed 20 February 2018.
26 'Deimos as Viewed by Mars Reconnaissance Orbiter'.
27 Quoted in Elizabeth Zubritsky, 'Mars' Moon Phobos is Slowly Falling Apart', NASA, 19 November 2015, www.nasa.gov.
28 Bevan Sharpless, 'Secular Accelerations in the Longitudes of the Satellites of Mars', *Astronomical Journal*, LI (1945), p. 185.
29 Zubritsky, 'Mars' Moon Phobos is Slowly Falling Apart'.
30 'Mars Moons', NASA, https://solarsystem.nasa.gov, accessed 14 February 2018.
31 O'Meara, 'The Demon Sprites of Mars'.
32 Alan MacRobert, 'Hunting the Moons of Mars', *Sky and Telescope*, LXXVI/3 (1988), p. 280.

9 Observing Mars

1 Stephen James O'Meara, 'Mars in Daylight, No Telescope Needed', *Astronomy* (February 2019).
2 A. Sánchez-Lavega et al., 'An Extremely High-altitude Plume Seen at Mars' Morning Terminator', *Nature*, 518 (26 February 2015), pp. 525–8. Alexandra Witze, 'Martian Mystery Cloud Defies Explanation', *Nature*, 16 February 2015, www.nature.com.
3 Nola Taylor, 'Mars Life Hunt: Could Basin Host Remains of an Ancient Biosphere?', *Space.com*, 1 April 2016, www.space.com.
4 P. E. Geiser and the HiRise Team, 'Persistent Surface Changes in Solis Lacus, Mars', 43rd Lunar and Planetary Science Conference, 2012, www.lpi.usra.edu, accessed 1 March 2019.

Bibliography

Abramov, Oleg, and Stephen J. Mojzsis, 'Thermal Effects of Impact Bombardments on Noachian Mars', *Earth and Planetary Science Letters*, 442 (15 May 2016), pp. 108–20

Achenbach, Joe, 'NASA's 1976 Viking Mission to Mars did all that was Hoped for – Except Find Martians', *Washington Post*, 18 June 2016

Adams, W. S., and C. E. St John, 'An Attempt to Detect Water-vapor and Oxygen Lines in the Spectrum of Mars with the Registering Microphotometer', *Publications of the Astronomical Society of the Pacific*, XXXVII/37 (1926), pp. 158–9

Ainsworth, Diane, 'Pathfinder Results Featured in This Week's Science Magazine', NASA, https://mars.nasa.gov, accessed 21 March 2018

'Ancient Mars Bombardment Likely Enhanced Life-supporting Habitat', *Science News*, 5 April 2016, www.sciencedaily.com

Antoniadi, E. M., *The Planet Mars*, trans. Patrick Moore (Chatham, 1975)

Bauer, Markus, 'Leaky Atmosphere Linked to Lightweight Planet', ESA, 8 February 2018, https://sci.esa.int

—, 'Shining Light on the Aurora of Mars', ESA, www.esa.int, accessed 20 March 2018

Bell, II, Edwin V, 'Phobos Project Information', NASA (2016)

Boyle, Alan, 'Curiosity Rover Sees Life-friendly Conditions in Ancient Mars Rock', NBC News, 12 March 2013, www.nbcnews.com

Brand, Stewart, 'Controversy is Rife on Mars: Interviewing Carl Sagan and Lynn Margulis', National Space Society, https://space.nss.org, accessed 20 February 2019

Bregin, Elana, 'The Identity of Difference: A Critical Study of Representations of the Bushmen', partial submission for award of MA, University of Natal, 1998

Brown, Dwayne, et al., 'Mars Phoenix Lander Finishes Successful Work on Red Planet', NASA, 10 November 2008, www.nasa.gov

'Buried Glaciers on Mars', ESA, www.esa.int, accessed 22 March 2018

Campbell, W. W., 'The Spectrum of Mars', *Publications of the Astronomical Society of the Pacific*, VI (1894), pp. 228–36
Campion, Nicholas, *Astrology and Cosmology in the World's Religions* (New York and London, 2012)
Cantril, Hadley, *The Invasion from Mars: A Study in the Psychology of Panic* (Piscataway, NJ, 1940)
Chadwick, Stephen Robert, and Martin Paiour-Smith, *The Great Canoes in the Sky: Starlore and Astronomy of the South Pacific* (New York and London, 2017)
Christensen, P. R., et al., 'Evidence for Magmatic Evolution and Diversity on Mars from Infrared Observations', *Nature*, 436 (28 July 2005), pp. 504–9
'The Complete Phobos 2 VSK Data Set', The Planetary Society, www.planetary.org, accessed 18 February 2018
The Concise Mythological Dictionary (London, 1963)
Cooke, Lacey, 'NASA is Sending its Inflatable Mars Ice Home into Space', NASA, 25 October 2017, https://inhabitat.com
Crossley, Robert, 'Mars and the Paranormal', *Science Fiction Studies*, XXXV/3 (2008), pp. 466–84
—, 'Percival Lowell and the History of Mars', *Massachusetts Review*, XLI/3 (2000), pp. 297–318
'Curiosity Mission Overview', NASA/Jet Propulsion Laboratory, https://mars.nasa.gov, accessed 10 December 2018
De Vorkin, David H., 'W. W. Campbell's Spectroscopic Study of the Martian Atmosphere', *Quarterly Journal of the Royal Astronomical Society*, XVIII (1977), pp. 38–50
'The Dead Planet', *New York Times*, 30 July 1965, as reported in 'On Mars: Exploration of the Red Planet, 1958–1978', www.hq.nasa.gov, accessed 28 February 2018
'Deep Space 2 Probes Silent', NASA, 6 December 1999, https://mars.nasa.gov
Deffree, Suzanne, '1st U.S. Satellite Attempt Fails, December 6, 1957', 6 December 2018, www.edn.com
'Deimos as Viewed by Mars Reconnaissance Orbiter', The Planetary Society, www.planetary.org, accessed 20 February 2018
Dolan, David Sutton, 'Percival Lowell: The Sage as Astronomer', PhD diss., University of Wollongong, 1992
Dunbar, Brian, 'The Day the Internet Stood Still', NASA, www.nasa.gov, accessed 12 March 2018
Dundas, Colin M., et al., 'Exposed Subsurface Ice Sheets in the Martian Mid-latitudes', *Science*, 359 (12 January 2018), pp. 199–201
Dyches, Preston, 'NASA Radar Finds Ice Age Record in Mars' Polar Cap', JPL, 26 May 2016, www.jpl.nasa.gov
'Earthworms Can Reproduce in Mars Soil Simulant', Wageningen University and Research, www.wur.nl, accessed 1 February 2018

'ExoMars Highlights Radiation Risk for Mars Astronauts, and Watches as Dust Storm Subsides', https://exploration.esa.int, 19 September 2018

'Explorer 1 Overview', NASA, www.nasa.gov, accessed 20 February 2019

'First Mars Express Gravity Results Plot Volcanic History', ESA, www.esa.int, accessed 22 March 2018

Flammarion, Camille, *Camille Flammarion's The Planet Mars*, trans. Patrick Moore, ed. William Sheehan (New York and London, 2015)

—, 'Mars, by the Latest Observations', *Popular Science*, IV (1873)

Flournoy, Théodore, *From India to the Planet Mars: A Study of a Case of Somnambulism* (New York and London, 1900)

'Found It! Ice on Mars!', NASA, https://science.nasa.gov, accessed 14 March 2018

Fox, Susannah, 'The Internet Circa 1998', The Pew Research Center, 21 June 2007, www.pewinternet.org

Frazier, Sarah, 'Real Martians: How to Protect Astronauts from Space Radiation on Mars', NASA, 7 August 2017, www.nasa.gov

Fuller, Robert Stevens, 'The Astronomy of the Kamilaroi and Euahlayi Peoples and Their Neighbours', MPhil thesis, Macquarie University, 2014

Galilei, Galileo, *Le Opere di Galileo Galilei*, ed. Antonio Favaro (Florence, 1890), vol. X

—, and Johannes Kepler, *The Sidereal Messenger of Galileo Galilei and a Part of the Preface to Kepler's Dioptrics, Containing the Original Account of Galileo's Astronomical Discoveries*, trans. and ed. Edward Stafford Carlos (London, 1880)

Garcia, Mark, ed., 'Space Food Bars Will Keep Orion Weight Off and Crew Weight On', NASA, 22 November 2016, www.nasa.gov

Geiser, P. E., and the HiRise Team, 'Persistent Surface Changes in Solis Lacus, Mars', 43rd Lunar and Planetary Science Conference, 2012, www.lpi.usra.edu, accessed 1 March 2019

Geissler, Paul, 'Three Decades of Martian Surface Changes', *Journal of Geophysical Research*, CX (2005), pp. 1–23

'Gemstone of the Year – Labeled', NASA, 28 October 2008, https://mars.nasa.gov

Gingerich, Owen, 'Johannes Kepler and the New Astronomy', *Quarterly Journal of the Royal Astronomical Society*, XIII (1972), pp. 346–73

—, 'The Satellites of Mars: Prediction and Discovery', *Journal for the History of Astronomy*, I (1970), pp. 109–15

Glasstone, Samuel, *The Book of Mars* (Washington, DC, 1968)

Glotch, T. D., et al., 'Spectral Properties of Phobos from the Mars Global Surveyor Thermal Emission Spectrometer: Evidence for Water and Carbonate', 46th Lunar and Planetary Science Conference: The Woodlands, TX, 2015

Gray, R. A., 'The Life and Work of Tychi Brahé', *Journal of the Royal Astronomical Society of Canada*, XVII (1923), p. 104

Greicius, Tony, ed., 'What Happened to Early Mars' Atmosphere? New Study Eliminates One Theory', NASA, 8 August 2017, www.nasa.gov

Gronstal, Aaron, 'Putting the Ethics into Planetary Protection', NASA, 13 August 2018, https://astrobiology.nasa.gov

Guarno, Ben, 'Stephen Hawking Calls for a Return to the Moon as Earth's Clock Runs Out', *Washington Post*, 21 June 2017, www.washingtonpost.com

Hall, Angelo, *An Astronomer's Wife: A Biography of Angeline Hall* (Baltimore, MD, 1908)

Hall, Asaph, 'The Discovery of the Satellites of Mars', *Monthly Notices of the Royal Astronomical Society*, XXXVIII (1878)

Hand, Eric, 'On Mars, Atmospheric Methane – A Sign of Life on Earth – Changes Mysteriously with the Seasons', *Science*, 3 January 2018, www.sciencemag.org

Harden, Mary, 'Odyssey Finds Water Ice in Abundance Under Mars' Surface', Jet Propulsion Laboratory, 28 May 2002, www.jpl.nasa.gov

'Hawai'i Space Exploration Analog and Simulation', Hi-Seas, https://hi-seas.org, accessed 25 January 2018

Herschel, William, 'On the Remarkable Appearances at the Polar Regions of the Planet Mars, the Inclination of its Axis, the Position of its Poles, and its Spheroidal Figure; with a Few Hints Relating to its real Diameter and Atmosphere', in *The Scientific Papers of Sir William Herschel* (London, 1912), vol. I

Hockey, Thomas, with Virginia Trimble and Thomas R. Williams, eds, *Biographical Encyclopedia of Astronomers* (New York, 2007)

Holmes, Charles Nevers, 'Nicolaus Copernicus', *Popular Astronomy*, XXIV (1916)

'Home Computer Access and Internet Use', Child Trends Databank, www.childtrends.org, accessed 12 September 2019

Huggins, William, 'On the Spectrum of Mars, with Some Remarks on the Colour of the Planet', *Monthly Notices of the Royal Astronomical Society*, XXVII (1867), pp. 178–81

'Human Needs: Sustaining Life During Exploration', NASA, 16 April 2007, www.nasa.gov

Joffe, Justin, 'NASA's Kilopower Project Will Bring Us to Mars and Beyond With Nuclear Power', *Outer Places*, 22 January 2018, www.outerplaces.com

Jones, Alexander, ed., *Ptolemy in Perspective: Use and Criticism of his Work from Antiquity to the Nineteenth Century* (New York, 2010)

'Journey to Mars Overview', NASA, www.nasa.gov, accessed 25 January 2018

Kasak, Enn, and Raul Veede, 'Understanding Planets in Ancient Mesopotamia', *Folklore*, 16 (2001), pp. 6–33, available at www.folklore.ee

Kostinsky, S., 'Sur les photographies des satellites de Mars', *Astronomische Nachrichten*, CLXXXVI (1909)

Krause, Henning, and Ralf Jaumann, 'Images of Mars Express' Closest ever Flyby at Phobos', ESA, 30 July 2008, www.dlr.de

Lakdawalla, Emily, 'Curiosity's Discoveries on Mars', *Sky and Telescope* (April 2017)

Launay, Françoise, *The Astronomer Jules Janssen: A Globetrotter of Celestial Physics*, trans. Storm Dunlop (New York and London, 2012)

Launius, Roger D., 'Sputnik and the Origins of the Space Age', NASA, https://history.nasa.gov, accessed 20 February 2019

Lewis, Richard S., *From Vinland to Mars: A Thousand Years of Exploration* (New York, 1978)

'Lockheed Martin Reveals New Details to its Mars Base Camp Vision', Lockheed Martin, 28 September 2017, https://news.lockheedmartin.com

Locklear, Mallory, 'NASA Tests Small Nuclear Reactor that Could Power a Habitat on Mars', NASA, 18 January 2018, www.engadget.com

Lowell, Percival, 'The Polar Snows', *Popular Astronomy*, II (1894)

McEwan, A. S., et al., 'Global Color Views of Mars', Abstracts of the 25th Lunar and Planetary Science Conference, held in Houston, Texas, 14–18 March 1994, p. 871

McKim, Richard, 'The Dust Storms of Mars', *Journal of the British Astronomical Association*, CVI (1996), pp. 185–200

—, 'Telescopic Martian Dust Storms: A Narrative and Catalogue', *Memoirs of the British Astronomical Association*, XLIV (1999), pp. 14–15

McNab, David, and James Younger, *The Planets* (New Haven, CT, 1999)

'Mariner 3', NASA, https://nssdc.gsfc.nasa.gov, accessed 27 February 2018

'Mariner 9', NASA, https://nssdc.gsfc.nasa.gov, accessed 1 March 2018

'Mariner 9 Anniversary/Landslides on Mars', NASA, www.jpl.nasa.gov, accessed 1 March 2018

Markley, Robert, *Dying Planet: Mars in Science and the Imagination* (Durham, NC, 2005)

'Mars 1', NASA, https://nssdc.gsfc.nasa.gov, accessed 27 February 2018

'Mars-96', www.russianspaceweb.com, accessed 27 February 2018

'Mars Base Camp', Lockheed Martin, https://lockheedmartin.com, accessed 22 January 2018

'Mars Dust Storm', ESA, 19 July 2018, www.esa.int

'Mars Express Confirms Methane in the Martian Atmosphere', ESA, 30 March 2004, www.esa.int

'Mars Express Detects Liquid Water Hidden Under Planet's South Pole', ESA, 25 July 2018, www.esa.int

'Mars Express Science Highlights: #4. Probing the Polar Regions', ESA, 3 June 2013, www.esa.int

'Mars Global Surveyor: Important Discoveries', NASA, https://mars.jpl.nasa.gov, accessed 13 March 2018

'Mars has Complex Volcanic Processes', NASA, https://themis.asu.edu, accessed 14 March 2018

'Mars Ice House', www.marsicehouse.com, accessed 25 January 2018

'Mars Moons', NASA, https://solarsystem.nasa.gov, accessed 14 February 2018

'Marsnik 1', NASA, https://nssdc.gsfc.nasa.gov, accessed 26 February 2018

'Mars Odyssey (Themis): Olivine-rich Rocks Point to Cold, Dry Martian Past', NASA, https://themis.asu.edu, accessed 14 March 2018
'Mars Orbiter Mission (MOM) Completes 4 Years in its Orbit', Indian Space Research Organisation, 24 September 2018, www.isro.gov.in
'Mars Pathfinder', NASA, https://mars.nasa.gov, accessed 21 March 2018
'Mars Pathfinder Winds Down After Phenomenal Mission', NASA, https://mars.jpl.nasa.gov, accessed 12 March 2018
'Mars Reconnaissance Orbiter: NASA Science Goals', Jet Propulsion Laboratory, www.nasa.gov, accessed 20 February 2019
'Mars Terraforming Not Possible Using Present-day Technology', NASA, 30 July 2018, www.nasa.gov
'Mars Water-ice Clouds Are Key to Odd Thermal Rhythm', Jet Propulsion Observatory, 12 June 2013, www.jpl.nasa.gov
'Martian Moons', NASA, https://history.nasa.gov, accessed 19 February 2018
'Martian Moons: Phobos', ESA, https://sci.esa.int, accessed 20 February 2018
'Melissa', ESA, www.esa.int, accessed 28 January 2018
Meltzer, Michael, 'When Biospheres Collide: A History of NASA's Planetary Protection Programs', NASA Publication SP-2011-4234 (2010)
Messier, Douglas, 'Will Russia End its Curse at Mars?', *Space Review*, 7 November 2011, www.thespacereview.com
'Missions to Mars', The Planetary Society, www.planetary.org, accessed 1 March 2018
Molaro, Paolo, 'Francesco Fontana and his *Astronomical* Telescope', *Journal of Astronomical History and Heritage*, XX/2 (2017), pp. 271–88
Mosher, Dave, 'Elon Musk says SpaceX is on Track to Launch People to Mars within 6 Years: Here's the Full Timeline of his Plans to Populate the Red Planet', *Business Insider*, 2 November 2018, www.businessinsider.com
'NASA Encounters the Perfect Storm for Science', NASA, 18 June 2018, https://mars.nasa.gov
'NASA's MAVEN Reveals Most of Mars' Atmosphere Was Lost to Space', NASA, 30 March 2017, www.nasa.gov
'A New Home on Mars: NASA Langley's Icy Concept for Living on the Red Planet', NASA, 29 December 2016, www.nasa.gov
Norris, R. P., and D. W. Hamacher, ''The Astronomy of Aboriginal Australia', *The Role of Astronomy in Society and Culture: Proceedings of the International Astronomical Union, IAU Symposium*, vol. CCLX (2011), pp. 39–47
Novaković, Bojan, 'Senenmut: An Ancient Egyptian Astronomer', *Publications of the Astronomical Observatory of Belgrade*, 85 (2008), pp. 19–23
O'Meara, Stephen James, 'The Demon Sprites of Mars', *Sky & Telescope*, CI/6 (2001), pp. 102–4
—, 'Mars in Daylight, No Telescope Needed', *Astronomy* (February 2019)

—, *Night Skies of Botswana: Suitable for all Stargazing in the Southern Hemisphere* (Capetown, 2019)

—, private communication and personal experiences with Botswana's Basarwa

—, 'Tales from the Pacific', *Sky and Telescope*, LXXII/1 (1986), p. 73

Pannekoek, Antonin, *A History of Astronomy* (New York, 1961)

Parker, R. A., 'Ancient Egyptian Astronomy', *Philosophical Transactions of the Royal Society of London*, CCLXXVI (1974), pp. 51–65

Pascu, D., S. Erard, W. Thuillot and V. Lainey, 'History of Telescopic Observations of the Martian Satellites', *Planetary and Space Science*, CII (2014), pp. 2–8

Perrotto, Trent, and Deb Schmid, 'Radiation Measured by NASA's Curiosity on Voyage to Mars has Implications for Future Human Missions', NASA, 30 May 2013, www.nasa.gov

'Phobos', JPL, https://mars.jpl.nasa.gov, accessed 20 February 2019

Piantadosi, Claude A., *The Biology of Human Survival: Life and Death in Extreme Environments* (Oxford, 2003)

Pickering, E. C., A. Searle and W. Upton, 'Photometric Observations: The Satellites of Mars', *Annals of the Harvard College Observatory*, XI (1879), pp. 226–38

Plotkin, Howard, 'William H. Pickering in Jamaica: Founding of Woodlawn and Studies of Mars', *Journal for the History of Astronomy*, XXI (1993), pp. 101–22

Powley, Kathryn, 'Getting Sick in Space on the Way to Mars', University of Melbourne, 5 October 2017, https://pursuit.unimelb.edu.au

Proctor, Richard A., *Other Worlds Than Our Own* (New York, 1871)

'Project Viking Fact Sheet', Jet Propulsion Laboratory, https://astro.if.ufrgs.br, accessed 12 March 2018

Pultarova, Tereza, 'Phobos-Grunt Probe Adds Chapter to Russian Space Disaster Series', *Space Safety Magazine*, 11 November 2011, www.spacesafetymagazine.com

'Radiation Exposure Comparisons with Mars Trip Calculation', NASA, 1 January 2019, www.nasa.gov

'Recycling for Moon, Mars and Beyond', NASA, www.nasa.gov, accessed 31 January 2018

Redd, Nola Taylor, 'Mars Volcanic Glass May Be Hotspot for Life', NBC News, 1 May 2012, www.nbcnews.com

'Report from Mars: Mariner IV, 1964–1965', NASA, www.scribd.com, accessed 1 February 2019

Richardson, Robert, *Exploring Mars* (New York, 1954)

Rosen, Edward, 'Commentariolus', introduction to *Nicholas Copernicus: Minor Works* (Warsaw and Cracow, 1985), https://copernicus.torun.pl

Rosenberg, Eli, 'In the Coldest Village on Earth, Eyelashes Freeze, Dinner is Frozen and Temperatures Sink to –88°F', *Washington Post*, 16 January 2018

Sagan, Carl, 'The Solar System', *Scientific American*, CCXXXIII (1975), pp. 22–31
—, Tobias Owen and H. J. Smith, *Planetary Atmospheres* (Dordrecht, 1971)
Sánchez-Lavega, A., et al., 'An Extremely High-altitude Plume Seen at Mars' Morning Terminator', *Nature*, 518 (26 February 2015), pp. 525–8
Schaff, Michael, 'Dispelling Myths and Highlighting History of the Heliocentric Model', *Physics Today*, LXI (2008)
'Searching for Life on Mars: Development of the Viking Gas Chromatograph Mass Spectrometer', NASA, https://appel.nasa.gov, accessed 20 February 2019
Selk, Avi, 'Stephen Hawking, Kris Jenner and Other Famous People with Plans to Send Humans to Mars', *Washington Post*, 22 June 2017, www.washingtonpost.com
Sharpless, Bevan, 'Secular Accelerations in the Longitudes of the Satellites of Mars', *Astronomical Journal*, LI (1945), pp. 185–6
Sheehan, William, *The Planet Mars* (Tucson, AZ, 1966)
—, and Stephen James O'Meara, *Mars: The Lure of the Red Planet* (Amherst, NY, 2001)
Shivni, Rashmi, 'NASA Twins Study Spots Thousands of Genes Toggling On and Off in Scott Kelly', PBS, 28 October 2017, www.pbs.org
Smith, Brad, 'Phobos Preliminary Results from Mariner 7', *Science*, CLXVIII (1970), pp. 828–30
Smith, S. M., et al., 'Benefits for Bone from Resistance Exercise and Nutrition in Long-duration Spaceflight: Evidence from Biochemistry and Densitometry', *Journal of Bone and Mineral Research*, XXVII/9 (2012), pp. 1896–1906
Smith, William, ed., *Dictionary of Greek and Roman Biography and Mythology*, 3 vols (London, 1846)
Sols, Stephen, 'Copernicus and the Church: What the History Books Don't Say', *Christian Science Monitor*, 19 February 2013, www.csmonitor.com
Space Handbook: Astronautics and Its Applications: 85th Congress, 2nd session. House of Representatives staff report, NASA, 27–30 December 1958, https://history.nasa.gov
'Space Launch System: Building America's New Rocket for Deep Space Exploration', NASA, www.nasa.gov, accessed 20 January 2018
'Spirit and Opportunity', NASA, https://mars.nasa.gov, accessed 15 March 2018
'Sputnik 22', NASA, https://nssdc.gsfc.nasa.gov, accessed 20 February 2019
Swift, Jonathan, *Gulliver's Travels* [1726] (London, 1892)
Taylor, Nola, 'Mars Life Hunt: Could Basin Host Remains of an Ancient Biosphere?', *Space.com*, 1 April 2016, www.space.com
Verseux, Cyril, et al., 'Sustainable Life Support on Mars – The Potential Roles of Cyanobacteria', *International Journal of Biology*, XV (2016)
'Viking 1 and 2', NASA, https://mars.nasa.gov, accessed 12 March 2018
'Viking Mission to Mars', NASA, www.jpl.nasa.gov, accessed 20 February 2019

Vogt, Yngve, 'World's Oldest Ritual Discovered: Worshipped the Python 70,000 Years Ago', *Apollon*, 1 February 2012, www.apollon.uio.no

'W. H. Pickering to E. C. Pickering', Harvard College Observatory Director's correspondence (Cambridge, MA, 1877)

Wall, Mike, 'Mars Soil May Be Toxic to Microbes', *Space.com*, 6 July 2017, www.space.com

—, 'Hibernating Astronauts May Be Key to Mars Colonization', *Space.com*, 30 August 2016, www.space.com

Wallace, Alfred R., *Man's Place in the Universe: A Study of the Results of Scientific Research in Relation to the Unity or Plurality of Worlds* (London, 1904)

Wallace, Tim, 'First Mission to Mars: Mariner 4's Special Place in History', *Cosmos Magazine*, 14 July 2017, https://cosmosmagazine.com

'Waste Not, Want Not, On The Road To Mars', ESA, www.esa.int, accessed 1 February 2018

Webster, Guy, 'Mars May Be Emerging from an Ice Age', Jet Propulsion Laboratory, www.jpl.nasa.gov, accessed 14 March 2018

—, 'NASA Phoenix Mars Lander Confirms Frozen Water', NASA, 20 June 2008, www.nasa.gov

—, 'NASA – Report Reveals Likely Causes of Mars Spacecraft Loss', NASA, 13 April 2007, www.nasa.gov

—, 'NASA Spacecraft Images Mars Moon in Color and in 3D', NASA, 14 July 2009, www.nasa.gov

—, Jennifer LaVista and Laurie Cantillo, 'Steep Slopes on Mars Reveal Structure of Buried Ice', NASA, 11 January 2018, https://mars.nasa.gov

Weir, Andy, 'If We're Serious about Going to Mars, We Need Artificial Gravity', 4 March 2014, www.space.com

Wells, H. G., *The War of the Worlds* (London, 1898)

Westfall, John, and William Sheehan, *Celestial Shadows: Eclipses, Transits, and Occultations* (New York, 2015)

Witze, Alexandra, 'Martian Mystery Cloud Defies Explanation', *Nature*, 16 February 2015, www.nature.com

Yallop, Jacqueline, 'Force of Nature: Picturing Ruskin's Landscape', www.guildofstgeorge.org.uk, accessed 21 January 2018

Yates, Naona, 'Millions Visit Mars – on the Internet', *Los Angeles Times*, 14 July 1997, https://articles.latimes.com

Zharkov, V. M., and A. V. Kozenko, 'Fobos I Deimos – sputniki Marsa', *Novoye v zhizni, nauke, tekhnike, seriya: kosmonavtika, astronomiya*, 1 (Moscow, 1985), pp. 1–64; trans. as 'Phobos and Deimos: Satellites of Mars', NASA Technical Memorandum TM-88016, www.researchgate.net, accessed 7 May 2019

Zubritsky, Elizabeth, 'Mars' Moon Phobos is Slowly Falling Apart', NASA, 19 November 2015, www.nasa.gov

Acknowledgements

I would like to thank my friend, fellow author and Mars historian William Sheehan for initiating the project and introducing me to Dr Peter Morris of Reaktion Books, who encouraged the production of this work for the publisher's Kosmos series, which helps readers to better understand the historical, contemporary and future developments of outer space exploration. Daniel W. E. Green and Michael Rudenko of Harvard were instrumental in making the Harvard College Observatory library and archives available to me during my research. At the project's start, I received invaluable insight and suggestions from Dr Peter Morris, who helped set the book's tone, so thank you. I would also like to thank my editor, Amy Salter, for her swift and efficient editing, and for helping to keep the project on track. Helping her was Reaktion's capable design and production team, especially Alexandru Ciobanu (Assistant to the Publisher). Finally, I would like to thank my loving wife Deborah for her valuable comments and suggestions and for keeping the fire burning during this literary journey.

Photo Acknowledgements

The author and publishers wish to express their thanks to the following sources of illustrative material and/or permission to reproduce it.

From *Astounding Science Fiction*, XLI/5 (July 1958): p. 152; from the *Atti della R. Accademia dei Lincei – Mem. Cl. sc. fis. ecc.* Serie 5^{a}, vol. VIII (1890): pp. 37, 41 (left); courtesy the author: pp. 8, 9, 10, 11, 16, 17, 19, 20 (top), 21, 22, 95 (right), 135, 161 (left-hand side of image), 180 (left-hand side of image), 184 (adapted from an Association of Lunar and Planetary Observers illustration); from Wilhelm Beer and J. H. Mädler, *Beiträge zur physischen Kenntniss der himmlischen Körper im Sonnensysteme . . .* (Weimar, 1841): p. 32; 'COSMOS: A PERSONAL VOYAGE'/ Druyan-Sagan Associates, Inc.: p. 73; photos ESA: pp. 124, 161 (right-hand side of image); ESA/ATG medialab: pp. 125, 126; ESA/DLR/FU Berlin: pp. 93, 98; ESA/DLR/FU Berlin / Bill Dunford: p. 92; ESA/DLR/FU Berlin (G. Neukum): pp. 94, 96, 99, 100, 101, 173; from Camille Flammarion, *La Planète Mars et ses conditions d'habitabilité*, vol. I (Paris, 1892): pp. 27, 28; from Théodore Flournoy, *From India to the Planet Mars: A Study of a Case of Somnambulism with Glossolalia* (New York and London, 1900): p. 45; from [Francesco Fontana], *Novae Cœlestium Terrestriumq[ue] rerum observationes . . . à Francesco Fontana* (Naples, 1646): p. 26; from Pietro Gassendo [Pierre Gassendi], *Tychonis Brahei, Equitis Dani, Astronomorum Coryphæi, Vita* (The Hague, 1655) p. 22; collection of Owen Gingerich: p. 22; from Angelo Hall, *An Astronomer's Wife: The Biography of Angeline Hall* (Baltimore, MD, 1908): p. 157; Harvard Map Collection Digital Library (Cambridge, MA): p. 38; from Edward S. Holden, 'The Lowell Observatory, In Arizona', *Publications of the Astronomical Society of the Pacific*, VI/36 (July 1894): p. 52; Phil James (University of Toledo), Steve Lee (University of Colorado), NASA: p. 186; from *Journal of Geophysical Research*, 22 November 2006: p. 88; photo JPL/NASA/STSCI: p. 182; Lockheed Martin: p. 140; Metropolitan Museum of Art (Open Access): p. 15; Elon Musk/SpaceX: p. 151; NASA: pp. 13 (right), 58, 63, 64, 67 (foot),

76, 91, 95, 97, 102, 131, 134, 141, 147, 167, 180 (right-hand side of image), 185; photos NASA/Clouds AO/SEArch: pp. 142, 143; NASA, ESA, and Z. Levay (STSCI): p. 178; photo NASA/Dimitri Gerondidakis: p. 146; photo NASA Goddard Space Flight Center: p. 122; photos NASA/JPL: pp. 62, 67 (top), 78, 81, 82, 87, 90; photos NASA/JPL/Arizona State University, R. Luk: p. 109; photos NASA/JPL/Caltech: pp. 66, 115, 116, 127; photos NASA/JPL-Caltech/ESA/DLR/FU Berlin/MSSS: p. 114; photo NASA/JPL-Caltech/MSSS: pp. 123, 181; photo NASA/JPL-Caltech/UA/USGS: p. 107; photos NASA/JPL-Caltech/University of Arizona: pp. 108, 110, 170; photos NASA/JPL / Emily Lackdawalla: pp. 165, 166; photo NASA/JPL/Malin Space Science Systems: p. 86; photos NASA/JPL/Piotr Masek: p. 164; photo NASA/JPL/University of Arizona: p. 169; photo NASA/JPL/USGS/Phil Stooke: p. 65; photo The New York Public Library: p. 20 (foot); *Popular Science Monthly*, LXXVIII/4 (April 1911): p. 18; LXXXVIII/2 (February 1916): p. 41 (right); from Richard A. Proctor, *Other Worlds than Ours: The Plurality of Worlds Studied under the Light of recent Scientific Researches* (New York, 1896): p. 35; courtesy SpaceX: p. 153; photo United States Naval Observatory Library, Washington, DC: p. 156; University of Colorado/NASA: p. 118; from H. G. Wells, *La Guerre des Mondes* (Brussels, 1906): p. 47; Charles Wolf, *Histoire de l'Observatoire de Paris de sa fondation a 1793* (Paris, 1902): p. 29.

Index

Page numbers in ***bold italics*** refer to illustrations